Metal Detecting

The Ultimate Guide to Metal Detecting

(Learning to Use Metal Detector to Unearth All Sorts of Interesting)

Leon Johnson

Published By **Kate Sanders**

Leon Johnson

Metal Detecting: The Ultimate Guide to Metal Detecting (Learning to Use Metal Detector to Unearth All Sorts of Interesting)

ISBN 978-1-7776010-6-5

No part of this guidebook shall be reproduced in any form without permission in writing from the publisher except in the case of brief quotations embodied in critical articles or reviews.

Legal & Disclaimer

The information contained in this book is not designed to replace or take the place of any form of medicine or professional medical advice. The information in this book has been provided for educational & entertainment purposes only.

The information contained in this book has been compiled from sources deemed reliable, and it is accurate to the best of the Author's knowledge; however, the Author cannot guarantee its accuracy and validity and cannot be held liable for any errors or omissions. Changes are periodically made to this book. You must consult your doctor or get professional medical advice before using any of the suggested remedies, techniques, or information in this book.

Table Of Contents

Chapter 1: The History Of Metal Detecting

The captivating pursuit of metallic detecting, as we understand it these days, has its roots within the technological expertise of the 19th century and the technological advances of the 20 th. Let's adventure once more in time and find out the fascinating statistics of this super interest.

The foundation of metal detection lies within the medical breakthroughs of the early 1800s. Renowned physicist James Clerk Maxwell's electromagnetic precept, which located the relationship amongst power and magnetism, and Michael Faraday's pioneering art work on induction, laid the concept for the improvement of metallic detection technology.

However, it wasn't till the late 19th century that the first metallic detector got here into lifestyles. In 1881, following an assassination attempt on President James Garfield, Alexander Graham Bell, famed for inventing the mobile phone, evolved a primitive metallic detector in an unsuccessful try to find out the bullet lodged in the president's body. Despite the failure of Bell's tool in this immoderate-profile software, his format marked the number one use of a device recognizably comparable to trendy metal detectors.

The idea of metallic detecting become then in large part dormant until the early twentieth century at the same time as Gerhard Fisher, an engineer and passionate outdoorsman, made a big leap beforehand. While running on navigation systems, Fisher found anomalies because of the presence of ore-bearing rocks, which brought about the development of the number one portable metal detector within the Nineteen Twenties.

The following few a long time found metallic detectors in the fundamental used for mineral prospecting and software application detection. However, the advent of World War II introduced a present day utility for the ones gadgets: mine detection. Military want spurred upgrades in generation, making the detectors more compact and inexperienced.

In the Sixties, with the creation of transistor generation, metallic detectors have become greater much less highly-priced and portable, essential to an growth in reputation among hobbyists. It modified into all through this era that steel detecting transitioned from a expert tool right right into a leisure interest, starting the door for lovers to find out hidden treasures.

The Nineteen Seventies and '80s witnessed a golden age of treasure looking, with awesome reveals stated round the arena. This era additionally noticed the rise of metallic-detecting golf equipment and companies, selling accountable detecting and fostering a revel in of community amongst fans.

Today, steel detecting is a famous interest worldwide. Advances in era have allowed detectors to be more touchy, discriminating, and user-pleasant. From unearthing ancient cash in European fields to discovering gold nuggets in the Australian outback, steel detecting constantly contributes to our information of facts and gives a interesting interest for those who partake.

The statistics of steel detecting is a tale of scientific discovery, technological innovation, and the now not feasible to resist human strength to find out the secrets and techniques and techniques of the past. As we use our steel detectors, we aren't simply trying to find treasures hidden beneath the floor—we're moreover taking part in a way of life rooted in facts, as captivating due to the fact the gadgets we're seeking. This wealthy past serves as a reminder that each beep and sign is a part of an ongoing narrative, a story wherein we, as metal-detecting lovers, play an lively characteristic. As we sweep our detectors all through the panorama, we're not most effective looking for relics of the past however additionally continuing a ancient journey it absolutely is as interesting due to the fact the treasures we unearth.

Understanding Metal Detectors: The Basics

If you have ever watched a metal detectorist sweeping their tool in the course of the floor, you have possibly perplexed how those charming machines in reality artwork. Understanding the generation in the back of metallic detectors is vital to without a doubt draw close to their capability and the strategies to optimize their normal overall performance.

In the handiest terms, a metallic detector makes use of electromagnetic fields to find out steel objects. Here's the manner it really works:

At the coronary coronary heart of a steel detector is an oscillator that produces an alternating modern. This current-day passes via a coil called the transmitter coil, which then creates an alternating magnetic challenge. When this magnetic area interacts with a steel object, it induces eddy currents— round electric powered currents that waft inside the item, efficaciously developing a secondary magnetic situation.

Now, right right here's wherein the magic of metallic detection is available in. The metallic detector additionally has a receiver coil, frequently organized in a manner that it does no longer have interaction with the transmitted magnetic region under regular times. However, on the same time because the secondary magnetic place—created thru the eddy currents within the steel item—comes into play, it interferes with the receiver coil. This interference is detected and amplified, triggering an audible signal or visible display that indicators the detectorist to the presence of a metal object underneath the ground.

Chapter 2: Choosing Your First Metal Detector

Taking the plunge into the area of metal detecting may be every exhilarating and barely daunting. One of the maximum important steps on this journey is deciding on your first metal detector. This bankruptcy is devoted to guiding you thru this way, discussing factors to bear in mind, breaking down one-of-a-kind kinds of detectors, or perhaps suggesting some famous fashions for novices.

The first hobby is your budget. Metal detectors can range from low cost access-diploma models to costly professional-grade ones. If you're clearly starting, making an investment in a mid-range model regularly movements an high-quality stability amongst charge and capability. While it is probably tempting to move for a top-tier version, remember that the most steeply-priced detector may not guarantee the exceptional well-known. It's your skills and patience that makes the difference.

Next, pick out out out what you're maximum interested in locating. Are you seeking out coins, relics, gold, or possibly a aggregate of the whole thing? Certain detectors excel specially regions. For instance, in case you're seeking to prospect for gold, you might want a excessive-frequency VLF or a PI detector, as gold tends to be determined in especially mineralized soils wherein the ones detectors excel.

Then, recall in which you will be detecting. If you propose to discover beaches or special regions close to water, you could want to ensure your detector is water resistant. If it's parks or fields, recollect a detector with an wonderful discrimination function to assist sift through not unusual trash dreams.

Comfort is some extraordinary element regularly not noted. Metal detecting can consist of numerous hours of on foot and sweeping the detector, so mild-weight and ergonomic designs can help make your detecting lessons more amusing.

Finally, don't forget the detector's adjustability and competencies. As a newbie, it's far crucial to discover a stability among simplicity and boom functionality. Many modern detectors provide man or woman-pleasant interfaces but furthermore permit for adjustments as you develop extra confident.

Regarding the sorts of metallic detectors for novices, there are three critical instructions to undergo in thoughts: Very Low Frequency (VLF), Pulse Induction (PI), and Multi-Frequency.

VLF detectors are the maximum not unusual, supplying properly sensitivity to various metals and the functionality to discriminate among different types. They're flexible, reasonably-priced, and best for beginners. A first-rate example of this form of detector is the Fisher F22, diagnosed for its climate resistance, wonderful discrimination, and simplicity of use.

PI detectors are lots plenty less not unusual for novices because of their higher charge and complexity. However, they'll be notable for gold searching and working in regions with high-floor mineralization.

Multi-frequency detectors are often greater luxurious however provide the power of searching for severa styles of dreams successfully. An instance is the Minelab Equinox 800, which gives simultaneous multi-frequency operation, water-evidence introduction, and advanced settings.

Remember, the "amazing" metal detector varies from person to character, counting on man or woman desires and options. It's important to investigate, look at evaluations, and in all likelihood be a part of a nearby metallic-detecting membership to attempt out specific fashions. Choosing your first metal detector is an exciting step for your metallic-detecting journey. Take a while to find a detector that suits you, and you'll be properly in your manner to uncovering the hidden treasures of the earth.

Techniques for Effective Metal Detecting

Unearthing hidden treasures via steel detecting includes greater than genuinely swinging a detector round. It requires strategic planning, an data of your tool, staying power, and, most significantly, approach. This financial ruin will manual you thru the essential strategies of powerful steel detecting, placing you up for worthwhile discoveries.

Before you start, understanding your metal detector is critical. All detectors, whether or not or no longer VLF, PI, or Multi-frequency, include a whole lot of settings that may be amazing-tuned to suit your detecting environment. Familiarize yourself collectively along with your detector's manual, and apprehend a manner to set up discrimination, sensitivity, ground stability, and running frequency.

One of the number one techniques to master is the sweep. The sweep is the way you float the detector searching for coil over the floor. Maintain a normal, sluggish, and controlled motion, maintaining the coil diploma and near the ground. An uneven or rushed sweep can also moreover bring about omitted targets or faux signs.

Learning how to pinpoint a goal because it must be is a few different critical talent. Once your detector gives a signal, you must slowly flow into the coil over the goal vicinity in a bypass sample (X-sample) to slim down the place. Some detectors have a pinpoint mode, which offers a stronger sign when you're straight away over the purpose.

Digging for the aim moreover requires technique. A hand-held pinpointer device can be reachable for zeroing in at the item as quick as a hole is made. When digging, make sure to exercising the "plug" approach to decrease damage to the floor. Cut a horseshoe or C-usual plug into the grass, fold it again to retrieve the purpose, then replace it.

Gridding is a manner used to systematically are looking for an area. By mentally (or bodily, with markers) dividing your seek place into a grid, you make certain that you cover the entire area thoroughly. This may be specially useful in open fields or beaches.

A vital a part of steel detecting is deciphering the indicators your detector gives. These can inform you plenty about the capacity aim: its duration, intensity, and even in all likelihood what form of metal it's far. Most detectors offer an example of purpose depth, and some provide target identity scales that categorize potential well-knownshows (like coins, rings, or trash).

Understanding soil conditions and adjusting your detector as a quit result is also vital. Highly mineralized soils, black sand, and saltwater can all have an effect on the overall overall performance of your detector. Using the ground balance feature to "song out" mineralization can considerably enhance detection intensity and accuracy.

Learning whilst to dig and whilst no longer to dig is an enjoy-based totally completely expertise. As a amateur, it is regularly high-quality to dig all desires to gain experience and apprehend your detector's indicators. Over time, you'll discover ways to decide profitable indicators from possibly junk.

Lastly, keep in mind the maximum vital hassle in steel-detecting success: staying strength. Metal detecting is not a quick-paced interest. It's about the fun of the quest and the pleasure of discovery, that can take time. Slow down, experience the process, and allow the story of every discover unfold at its very non-public pace. Effective metal detecting is as loads an artwork as it is a technology, combining technical expertise with intuition and patience. Practice those strategies, and you may be to your way to successful detecting and fascinating finds.

Chapter 3: Researching And Identifying Potential Locations

With the fundamentals of metallic detecting below your belt, you are one step in the direction of your first interesting discovery. But in advance than you begin swinging your steel detector, it is important to understand the importance of region in steel detecting. This monetary catastrophe will guide you in gaining knowledge of and figuring out capability web sites that might be teeming with hidden treasures.

Research is the cornerstone of a success metal detecting. It gadgets the stage for in which to appearance and what you'll possibly discover. Start your research with community ancient data. Libraries, county courthouses, or nearby historical societies can provide treasured resources like vintage maps, belongings deeds, city plans, or possibly diaries and newspapers. Online resources, which consist of virtual documents and ancient aerial snap shots websites, can also provide a wealth of statistics.

Old maps are in particular useful for figuring out historical net sites that could not exist, like abandoned homesteads, antique faculties, or former amassing places. Comparing old maps with modern maps can spotlight adjustments inside the landscape and thing you in the direction of promising websites.

Another effective method is to end up aware of places in which humans have spent time inside the beyond. Think about in which humans could possibly have congregated, executed, worked, or traveled. This may additionally need to encompass parks, fairgrounds, picnic areas, fields, antique paths, or near our our bodies of water.

However, figuring out a functionality internet website is truely the start. Before you get your detector out, it's critical to affirm the legality and accessibility of the internet site online. Always recognize private property and are looking for permission from the landowner in advance than metallic detecting.

For public lands, test with nearby or u.S. Of the united states rules, as metal detecting laws can vary notably.

Geographical elements also can play a large function for your location choice. Some areas are manifestly richer in positive metals because of geological formations and mineral deposits. For instance, gold is regularly discovered near quartz veins in vintage volcanic regions.

Beaches and rivers may be exquisite locations due to their recognition for leisure sports activities sports, and the non-prevent moving of sand or water can find previously buried gadgets. However, preserve in mind that saltwater seashores or mineralized soils can require unique detector settings or device.

Once you have got determined on a vicinity, plan your detecting method. Begin via way of doing a quick take a look at of the location to perceive "warm spots" with higher concentrations of goals. From there, you can start a extra thorough are looking for, the usage of a systematic approach like gridding to ensure you cover the region absolutely.

As you advantage experience, you may make bigger a "sixth enjoy" for properly places. Every find out, no matter how insignificant it would seem, gives a bit of the historical puzzle of the internet web page. Noting and facts the ones patterns can manual your future detecting efforts.

Remember, metallic detecting is a adventure of discovery, now not most effective for the items you discover but for the ancient context and memories they represent. Researching and figuring out capability locations shape the muse of this adventure, helping rework your interest right proper into a exciting quest into the beyond.

Metal Detecting Etiquette and Ethics

While steel detecting can be an interesting journey into data and geology, the amusing of discovery have to typically be tempered through understand for the beyond and the surroundings. This financial ruin delves into the essential hassle don't forget of etiquette and ethics in metallic detecting, offering you with a guiding principle on a way to treasure hunt responsibly.

First and number one, metallic detecting have to constantly be a criminal hobby. Always are in search of for permission to stumble upon on private lands and public lands, and make yourself familiar with nearby, u . S . A ., and federal criminal suggestions that govern steel detecting. Some regions, which includes countrywide parks or ancient internet sites, may additionally moreover moreover have restrictions or outright prohibitions on metal detecting.

Respecting the belongings and rights of others is an critical part of metallic-detecting etiquette. Leave gates as you discover them, keep away from unfavorable plants or disrupting herbal world, and continuously fill in any holes you dig. Leaving a website as you positioned it, or better, is thought in the detecting community because the "Code of the Hole," and following it demonstrates admire for the land and the hobby.

Artifact recovery is a particularly sensitive vicinity. In many jurisdictions, extensive historic or cultural artifacts cannot legally be eliminated or can also require reporting to authorities. When you find an artifact, it's miles critical to file its region correctly, as this records can offer vital context for archaeologists and historians.

Respect for records moreover extends to the treatment of unearths. The urge to easy or restore gadgets can be robust, however incorrect coping with can doubtlessly damage or damage historic devices. Preserve the object as quality you can in its determined country until you could are looking for recommendation on its right treatment.

Sharing your discoveries is one of the joys of steel detecting, but preserve in thoughts the capability outcomes. Revealing the locations of your finds can positioned web sites at chance, in particular in the occasion that they turn out to be dreams for unscrupulous detectorists or looters. Always be cautious about the statistics you percentage publicly, especially online.

Metal detecting moreover has a characteristic to play in environmental stewardship. As you look for treasures, you can assist smooth up the environment through way of doing away with trash and clutter you come upon. Not handiest does this improve the net internet page for everybody, however it additionally makes your detecting more powerful with the aid of lowering the style of trash goals.

Inclusion and understand for others are also vital. Metal detecting is a diverse hobby loved through human beings of every age and backgrounds. Encourage inclusivity via respecting exclusive detectorists, providing assist or recommendation to novices, and being a effective consultant of the interest.

Ultimately, steel detecting is ready more than simply the pursuit of treasure. It's approximately connecting with records, exploring the herbal international, and being part of a network of like-minded fanatics. By adhering to those standards of etiquette and ethics, you can make certain that steel detecting stays a cherished and sustainable hobby for generations to come again back.

Uncovering Finds: From Coins to Relics

The delight of metal detecting lies inside the suspense of not knowing what your next discover can be. From cash and jewelry to relics and artifacts, each discovery consists of its very very very own story, presenting a glimpse into the beyond. This bankruptcy introduces you to the style of famous you could assume and offers guidelines to maximize your achievement in unearthing those buried treasures.

The most not unusual reveals in steel detecting are cash. Coins are valuable not satisfactory for their cloth certainly really worth but also for their ancient and numismatic price. They can be placed almost anywhere, from rural fields to metropolis parks. Over the years, countless coins have been out of place or deliberately buried, and their long lasting metallic composition method they frequently stay on within the floor for masses or perhaps hundreds of years.

Jewelry is every different famous beauty of well-known. Rings, necklaces, brooches, and different ornaments frequently maintain more than genuinely cloth fee; they will be non-public gadgets that offer a direct connection to the individuals who misplaced them. Moreover, earrings devices can once in a while be valuable in phrases of precious metallic or gemstone content cloth.

Chapter 4: Cleaning And Preserving Your Finds

Congratulations! You've made a discovery along aspect your metallic detector. Now comes the vital step of making sure the longevity of your discover through appropriate cleansing and protection. It's vital to cope with your well-knownshows with care, as fallacious cleansing methods can damage your treasures or maybe lower their charge.

To begin with, it's far essential to correctly pick out out your locate to recognize the fabric it is made from, as cleansing techniques can vary considerably. The next steps will manual you via cleaning and retaining reveals of various sorts.

Coins: Rinse dirt and unfastened debris off with smooth water. Do now not rub or scrub, as this can scratch the coin. It's typically extraordinary to head away cash as decided. Cleaning may additionally lower the numismatic charge. For historic cash or those manufactured from precious metals, are looking for for advice from a professional in advance than any cleaning try.

Jewelry: Use mild cleaning cleaning soap and water for cleaning. A smooth toothbrush can help to cast off dust in crevices. Do no longer try and use any harsh chemicals or abrasive substances, and normally dry very well after cleansing.

Relics: These require careful dealing with. If you find relics made from iron, a way called electrolysis may be used for cleansing. However, this device calls for device and information. If you are uncertain, don't forget consulting with a expert conservator.

Artifacts: Remember, anything which could have archaeological significance want to no longer be cleaned until it has been mentioned and tested via experts.

Beyond cleansing, maintaining your reveals is similarly vital. This often includes storing your gadgets in a dry, temperature-controlled surroundings. Acid-loose materials need to be used for garage, as ultra-modern materials can also lead to corrosion or extraordinary harm over the years.

For delicate objects, it's miles fine to speak over with a conservation professional. Museums and universities regularly have conservation departments which can offer recommendation or services. Remember, the last purpose is to maintain the item for future generations to have a have a examine and recognize, so take it gradual and do the right element.

When considering the cleaning and protection of your well-knownshows, commonly hold in thoughts the ethical issues of metal detecting. Remember that a few objects, particularly the ones of historic or archaeological significance, need to be stated to network government or archaeological societies in advance than any cleaning attempt.

The pleasure of discovery that consists of metallic detecting goes hand in hand with the duty of safety. By being worried on your well-known because it must be, you aren't really maintaining artifacts; you are additionally maintaining the stories and histories they hold. Happy cleansing, preserving, and, most importantly, discovering!

Identifying and Valuing Your Finds

Metal detecting is not handiest about discovering hidden treasures below the ground; it's also approximately appreciating and expertise what you've got located. The method of identifying and valuing your reveals can be as thrilling because of the reality the hunt itself. In this bankruptcy, we delve into how you can grow to be aware about your discoveries and, if possible, verify their in reality worth.

Identifying Your Finds:

When you find out an object, the number one task is to choose out out it. This calls for a aggregate of observational abilties, research, and, once in a while, expert recommendation. Here's a elegant technique you may use:

a. Carefully check the object: Note its period, form, coloration, fabric, and any unique features or markings. These information may be clues to the item's identity and foundation.

b. Conduct research: Use reference books, on line databases, and boards devoted to steel detecting. They often have pics and outlines that will let you understand your find out.

c. Consult with experts: If you are having difficulty identifying an object, keep in mind consulting an professional. Museums, historical societies, or skilled individuals of steel-detecting golf equipment can regularly provide valuable insights.

Valuing Your Finds:

Valuing a discover isn't always straightforward, specifically for ancient or uncommon gadgets. However, there are techniques you could rent to get an approximate concept of an item's really worth:

a. Market evaluation: Check online public sale websites or collector's markets for comparable devices and be conscious what they're selling for. This will give you a hard concept of its modern-day market fee.

b. Professional appraisal: For precious or uncommon devices, don't forget seeking out a expert appraisal. Appraisers are skilled in evaluating and pricing gadgets based totally on their scenario, rarity, and marketplace call for.

c. Historical fee: Remember, not all charge is economic. Items with historic importance often hold an intrinsic fee that transcends any marketplace charge. They can be treasured domestic home windows into the past, telling memories of our shared human history.

While identifying and verifying your finds, bear in mind the ethical guidelines in metallic detecting. If you come across items of capability archaeological or historical importance, it is important to report them to the applicable government or close by archaeological societies.

Metal detecting is as loads approximately unearthing reminiscences from the past as it's miles approximately finding precious treasures. As you pick out out out and rate your exhibits, you're actively taking part in the interesting quest of piecing collectively our shared human history. So, as you experience the fun of discovery, commonly bear in mind the profound characteristic you play as a custodian of these portions of the beyond.

Law and Regulations: Rights and Responsibilities

The international of metal detecting is fraught with attraction and pleasure. However, every detectorist ought to moreover recognize that this interest is regulated thru way of felony pointers and recommendations, a number of which range hugely depending on geographic area. Adhering to the ones legal suggestions is important, now not surely to preserve the legality of your sports sports but additionally to uphold the ethics of treasure looking and respect for our shared cultural background.

Permission to Detect:

The first rule in metallic detecting is always to gain permission in advance than detecting on personal land. It's important to establish an facts with landowners approximately the division of any capability famous or to at least inform them about your sports sports. Don't assume that open fields, antique homes, or deserted houses are free for detecting. These places although belong to a person, and unauthorized detecting may be taken into consideration trespassing.

Laws Regarding Finds:

The law varies notably in phrases of who owns the exhibits. In many countries, collectively with america, the finder can preserve what they find out. In comparison, within the United Kingdom, the Treasure Act

(1996) calls for the reporting of fantastic classes of devices to the nearby coroner inner 14 days, who will then determine on possession and functionality museum hobby.

Archaeological Sites:

Detecting on registered archaeological net sites is precisely prohibited in many nations. Such web sites are regularly home to traditionally massive gadgets that should be left to professional archaeologists to excavate and maintain.

Beaches and Waterways:

Coastal areas, rivers, and lakes can also have unique prison guidelines for metallic detecting, in particular in terms of capability shipwreck websites. In a few regions, you could need a allow to hit upon in those locations. Always check with neighborhood tips in advance than you proceed.

Parks and Protected Areas:

Public parks, nature reserves, and protected landscapes often have policies or bans on metallic detecting to maintain the herbal and cultural environment. Always recognize such places and cling to the policies set.

The Code of Ethics:

Beyond felony recommendations and regulations, a self-imposed code of ethics is determined through manner of maximum responsible detectorists. This includes gaining permission, reporting big finds, no longer detecting on covered web web sites, respecting the geographical location, warding off disturbing plants and fauna, and refilling all holes.

Remember, it's miles essential to analyze the right legal recommendations and rules of your america and even your neighborhood location earlier than you begin steel detecting. This lets in ensure that your interest remains a jail and respectful pursuit of data. Following the regulation no longer most effective protects you but moreover protects our shared cultural background. Metal detecting may be a nice hobby, starting up a global of historic discovery. But with this thrill comes the obligation to ensure that our pursuit of the past respects the crook tips of the triumphing.

Chapter 5: Metal Detecting And Archaeology

Metal detecting may be a doorway to information, permitting us to unearth relics of the beyond hidden under our toes. However, it is essential to understand that this exciting interest intersects with the sphere of archaeology in complicated strategies.

Archaeology is the scientific have a have a examine of human records through the excavation of historic web web sites and the assessment of observed artifacts. Metal detecting, in assessment, is frequently driven through the use of a passion for records and the thrill of discovery. The anxiety amongst those fields arises from differing strategies to the coping with of artifacts.

Context is Key:

To an archaeologist, the fee of an artifact lies not simplest inside the object itself but furthermore in its context. This context includes its region, depth, and surrounding gadgets, all of which could provide profitable clues about beyond human sports activities. Therefore, on the same time as an artifact is removed by a detectorist without professional archaeological documentation, critical statistics approximately our facts can be misplaced.

Responsible Detecting:

Responsible steel detecting is ready putting a balance most of the pride of discovery and admire for our shared statistics. Always make certain you have got permission to search around and cling to the prison pointers and tips of your locality. If you believe you've got found an item of historical importance, it's far vital to record it to the appropriate authorities, who can examine and file the discover professionally.

Portable Antiquities Schemes:

In reaction to the upward push of metal detecting, some worldwide locations have installation applications to encourage voluntary reporting of well-knownshows via the usage of manner of the overall public. The UK's Portable Antiquities Scheme is a first-rate instance, permitting detectorists to percentage their discoveries with archaeologists who report the unearths and their facts, contributing to our collective knowledge of the past.

Collaborative Ventures:

There's growing reputation of the ability for collaboration amongst metallic detectorists and archaeologists. Cooperative ventures, in which detectorists paintings under archaeological supervision, can bring about extensive discoveries. These partnerships integrate the detectorists' abilities and neighborhood expertise with the archaeologists' information in excavating and decoding web sites.

Educational Opportunities:

As a detectorist, do not forget taking element in archaeology workshops or publications. This no longer simplest equips you with a better facts of the medical technique to historical well-knownshows but moreover lets in foster a better relationship among the metallic-detecting network and the archaeological global.

The worlds of metallic detecting and archaeology want not be at odds. Both are advocated through a love of facts and the joys of discovery. By education responsible detecting, reporting famous appropriately, and in search of possibilities for collaboration and training, detectorists can play a essential function in uncovering and keeping our shared beyond. The balance is touchy, however with care and recognize, it's totally viable to launch the mysteries of the beyond at the same time as retaining the treasures of our facts for future generations.

The Role of Geography and Geology in Metal Detecting

One of the excellent thrills of steel detecting is the unpredictability. Every beep can hold the ability for a hidden treasure. However, knowledge of geography and geology can rework your revel in from a sport of danger right into a greater calculated pursuit. This monetary break explores the feature of those fields in growing the achievement and enjoyment of your metal-detecting exploits.

Geography:

The term' geography' originates from the Greek' geographic,' which means that that 'earth description.' As metal detectorists, we engage in our very non-public form of 'earth description,' surveying the landscape for potential websites. An statistics of geographical ideas which incorporates human settlement styles, historic occasions, and land use can provide important clues approximately in which to appearance.

For example, vintage maps can suggest former buildings, pathways, and concern boundaries, often now invisible at the floor. Similarly, regions close to rivers, crossroads, or one of a kind herbal meeting factors can yield fruitful searches, as those had been frequently gathering locations for past societies.

Geology:

Geology is the take a look at of the Earth's strong materials and the strategies that form them. From a metallic detecting attitude, records the number one requirements of geology can be beneficial.

Different styles of soil and rock will have an effect on how without hassle a metal detector can penetrate the floor and stumble upon buried gadgets. For example, especially mineralized soils can cause interference inside the detector's signs, at the same time as sandy or loamy soils are normally less hard to art work with.

Ground Penetrating Radar (GPR):

For intense enthusiasts, technology like Ground Penetrating Radar can offer a more genuine facts of what lies underneath the surface. GPR works via sending high-frequency radio waves into the ground and measuring the reflected signs. This data can help pick out out out buried objects or systems and determine floor conditions in advance than you begin detecting them.

Be Aware of the Environment:

Lastly, but importantly, popularity of geographical and geological situations must also tell your technique to the environment. Avoid causing any erosion or disturbance to herbal capabilities. In sensitive regions, at the facet of seashore dunes or riverbanks, metallic detecting would probably cause extra damage than it's far in reality well worth.

By making use of expertise of geography and geology, you could expand extra strategic detecting plans, boom your opportunities of a achievement famous, and make sure you are being concerned for the environments in that you hit upon. Like the layers of the earth, the layers of statistics are ready to be explored. Let's preserve this exciting adventure into the location of metal detecting with an appreciation for the geography and geology that office work the very diploma of our treasure-searching adventures.

Adventures in Metal Detecting: Tales from the Field

Every beep of a metal detector holds the potential for an incredible discover, a totally particular tale from the past waiting to be informed. It's those exciting moments that regularly make the hobby of metallic detecting so addictive. In this bankruptcy, we're going to recount a few excellent testimonies from the world, each supplying each concept and schooling in your non-public treasure-looking adventures.

The Staffordshire Hoard:

One of the most well-known metallic-detecting discoveries is the Staffordshire Hoard. Found thru Terry Herbert in a subject in Staffordshire, England, in 2009, this treasure trove of 7th-century Anglo-Saxon gold and silver artifacts is valued at approximately £three.285 million. This discovery underscores the significance of staying power and thoroughness; Herbert modified into reportedly looking the sphere for 5 days in advance than stumbling upon this historic find.

The Ringlemere Cup:

In 2001, metallic detectorist Cliff Bradshaw decided the Ringlemere Gold Cup in a field in Kent, England. This Bronze Age vessel, dating from among 1700 and 1500 BC, is one in every of exceptional seven recounted "unstressed" gold cups from the duration. The cup, which modified into sadly overwhelmed with the resource of a plow in contemporary instances, offered at public sale to the British Museum for £270,000. This serves as a reminder that the rate of a locate frequently lies in its ancient importance, now not its look.

The Boot of Cortez:

Across the Atlantic in Mexico, in 1989, a hobbyist the usage of an inexpensive steel detector decided the biggest gold nugget currently said inside the Western Hemisphere. The so-referred to as "Boot of Cortez," weighing 389.Four troy ounces (about 26.6 pounds or 12.1 kilograms), supplied at auction for over $1.Five million in 2008. The lesson? Even the most number one device can result in super discoveries in case you're in the right place on the proper time.

The Mojave Nugget:

In 1977, Ty Paulsen located a nugget referred to as the Mojave Nugget within the Stringer district of California using a metallic detector. Weighing about 156 ounces. (approximately 10 kilos or four.Four kilograms), it is considered one of the most important nuggets positioned within the area. His discover underscores the capacity of metal detecting in gold-rich regions.

The Vale of York Hoard:

Discovered via father and son David and Andrew Whelan in North Yorkshire, England, in 2007, the Vale of York Hoard contained over 600 silver cash and unique Viking-generation artifacts. Valued at £1,082,000, the hoard changed into purchased together thru each the British Museum and the York Museums Trust. Their story highlights the pleasure of sharing the interest with loved ones, growing memorable reviews together.

These memories offer a glimpse into the exhilarating possibilities that lie beneath our ft. They illustrate how ordinary humans, equipped with metal detectors and a enjoy of journey, have unearthed high-quality treasures, rewriting history and incomes an area inside the annals of steel detecting lore. As you embark in your very very personal treasure-searching adventure, who's aware of? The subsequent awesome discovery can be high-quality a beep away.

Chapter 6: The World Of Underwater Metal Detecting

A sunken supply's hidden treasure, a misplaced piece of jewellery on a beach, or relics from an historic civilization submerged in a riverbed—those are a number of the charming famous that sit up for folks who challenge into the world of underwater metallic detecting. This branch of the hobby offers a completely specific blend of treasure searching and scuba diving or snorkeling, making it an journey in every enjoy of the word.

Tools of the Trade:

Underwater steel detecting calls for specialised gadget. Most importantly, your steel detector need to be water-evidence and capable of functioning well on the intensity you want to look. Some fashions are designed particularly for underwater use, providing insulated coils, water-proof housings, and headphones for underwater audio symptoms.

Just as important due to the fact the detector itself are diving or snorkeling equipment, depending at the depth of the water you are exploring. Always prioritize protection at the same time as diving—in no way flow by myself, test your system earlier than every dive, and strictly adhere to stable diving practices.

Where to Search:

Underwater metal detecting may be achieved in numerous our our bodies of water, each providing specific opportunities. Rivers and lakes, especially near antique settlements, can yield historic artifacts. Beaches and swimming areas are regularly pinnacle spots for finding out of place modern-day-day devices, like rings or cash. Underwater caves and shipwrecks, while requiring advanced diving abilities, can maintain interesting treasures from the beyond.

Famous Underwater Finds:

The Atocha Motherlode, decided through Mel Fisher in 1985 after a 16-12 months are looking for, is one of the most fantastic underwater well-knownshows. The Spanish galleon Nuestra Señora de Atocha sank near the coast of Florida in 1622, carrying a massive treasure of gold, silver, and valuable gems. Today, the haul is well worth over $400 million.

In 2015, underwater detectorists in Israel decided more than 2,000 gold cash from the Fatimid Caliphate (10th–twelfth centuries) in what is referred to as the Caesarea Gold Treasure. This serves as an wonderful example of the manner underwater metal detecting can display massive ancient artifacts.

Challenges and Ethics:

Underwater steel detecting brings its very personal set of stressful situations. Visibility may be limited, robust currents can pose risks, and pinpointing a goal can be more difficult underwater. These factors make this form of detecting more worrying however also more profitable for plenty hobbyists.

As with all varieties of metallic detecting, ethics must guide your movements. Any suspected historic or archaeological reveals must be said to the relevant government. In many regions, casting off artifacts from shipwrecks or included underwater websites is illegal. Be effective to understand the crook tips and policies to your region.

Venturing under the water's ground opens a brand new international of functionality discoveries. The sea, rivers, and lakes keep endless recollections from our beyond— testimonies that you, as an underwater detectorist, have the satisfactory possibility to maintain to the surface. Embarking on this underwater journey, you could discover that the magnetic appeal of metallic detecting runs even deeper than you imagined.

Gold Prospecting: A Specialized Pursuit

Since time immemorial, gold has captivated the human creativeness. Its gleam has ignited infinite expeditions, sparked legends, or perhaps fueled complete gold rushes. In the area of metal detecting, gold prospecting is a specialised pursuit, an adventure uniquely its very personal.

Gold Prospecting and Metal Detecting:

Prospecting for gold is not pretty like searching for cash, relics, or distinct steel gadgets. Gold regularly is to be had in small flakes or nuggets, which can with out hassle be disregarded with the useful resource of the use of a standard-reason metallic detector. Gold-precise detectors are designed to deal with pretty mineralized soil situations frequently discovered in gold-bearing areas, and they may sense even tiny fragments of gold at thoughts-blowing depths.

Choosing the Right Detector:

Gold prospecting detectors commonly function at higher frequencies to come across small gold nuggets, and that they consist of floor-balancing functions to deal with mineralized soils. Pulse induction (PI) detectors can be in particular powerful for gold prospecting, presenting superior intensity in as a substitute mineralized ground conditions. However, they'll be regularly extra costly than very low frequency (VLF) detectors, which additionally can be powerful and offer better discrimination capabilities.

Where to Prospect:

Not all ground is created identical in terms of gold prospecting. Gold is often determined in places with a records of gold mining. Riverbanks, appeared goldfields, and antique mines can be promising locations to appearance. Gold has a dishonest to gather in low-stress areas in rivers, which embody internal bends and behind massive boulders.

In many nations, prospecting rights or lets in can be had to look for gold. Always make certain that you are prospecting legally and ethically, respecting land possession and environmental policies.

Panning and Sluicing:

In addition to steel detecting, gold prospectors regularly use traditional techniques like panning and sluicing to separate gold from soil and gravel. A clean gold pan, swirled with water, can trap heavy gold particles, on the equal time as a sluice makes use of flowing water to capture gold at the equal time as letting lighter cloth wash away. These techniques can be mainly useful for processing soil from riverbeds or dry-washing soil from wasteland regions wherein water is scarce.

Notable Gold Discoveries:

There are many testimonies of detectorists putting gold. In 1980, the Mojave Nugget, weighing a whopping 156 oz..., turn out to be placed in California with a metal detector. And in 1989, a gold nugget named "Hand of Faith," weighing nearly 875 oz.., changed into decided through way of a hobbyist in Australia, proving that fantastic exhibits are absolutely viable.

While the dreams of fortune might also moreover moreover dance in the prospector's thoughts, keep in thoughts that gold prospecting is as a whole lot approximately the adventure as it's miles about the potential payout. The thrill of the hunt, the beauty of nature, and the satisfaction of unearthing a tangible link to geological records are rewards in and of themselves. Equipped with a chunk of facts, a sprint of staying electricity, and a wholesome dose of perseverance, you're ready to enroll in the ranks of gold prospectors in a quest as vintage as civilization itself.

Joining the Metal Detecting Community: Clubs and Organizations

Metal detecting can also look like a solitary pursuit, however there may be a thriving community of fans to be had who percentage your passion. Detectorists are related through manner of manner of the common thread of hobby, a revel in of adventure, and a mutual recognize for information. Joining a metal detecting club or corporation is an wonderful way to connect to fellow detectorists, look at from their reviews, proportion your finds, and take part in prepared hunts.

Why Join a Metal Detecting Club?

Metal detecting golf equipment provide an array of advantages. They host organization hunts that offer camaraderie and a pleasing competitive spirit. They offer a platform for sharing information and tips approximately device, strategies, and locations. They can also prepare traveller audio gadget, schooling periods, or perhaps treasure hunts. Often, similarly they recommend for the rights of detectorists, liaising with landowners and authorities. Perhaps most importantly, they carry together a community of like-minded individuals who understand the amusing of the quest.

Finding a Club:

There are severa metallic detecting golf equipment across the globe, each with its very very personal precise taste. Some cater to novices, others to experts; a few focus on historical well-knownshows, others on gold prospecting or seashore detecting. Some golf equipment have a network recognition, even as others are countrywide or international. You also can pick out a club based totally on its hobby, its member blessings, its region, or simply its network spirit.

In the us, the Federation of Metal Detector and Archaeological Clubs (FMDAC) is a famous corporation. In the United Kingdom, the National Council for Metal Detecting (NCMD) serves a similar function. Many countries have their personal national companies, and there are limitless local and close by clubs.

Online Communities:

In this digital age, you could additionally connect to the metal-detecting community on line. Internet boards, social media groups, and on-line groups provide an on hand manner to proportion reviews, ask questions, and examine from others, no matter your geographical vicinity. Websites including TreasureNet, Detectorist, and Metal Detecting Forum have masses of lively users sharing their well-knownshows, memories, and advice.

Organized Hunts and Rallies:

Many clubs and companies preserve prepared hunts or rallies, which can be exciting occasions for detectorists. These activities offer an opportunity to look new regions, compete in friendly competitions, and share reveals with a bigger agency. They can be day activities, weekend-extended, or even week-prolonged rallies in metallic-detecting hotspots.

Advocacy:

Joining a club is also a way to make contributions to the network's advocacy efforts. Clubs often art work to guard detectorists' rights, negotiate get admission to to net sites, and promote ethical detecting practices. Through a united voice, the community can more correctly have interaction with landowners, archaeologists, legislators, and the overall public.

Joining the metallic-detecting network enriches the hobby by way of way of supplying a social element, a studying platform, and a collective voice. Whether it is sharing the a laugh of a huge discover, studying from the stories of others, or reputation collectively to guard get right of access to to ancient internet sites, there may be loads to be obtained from being part of this colourful network. As you delve deeper into metal detecting, bear in thoughts the fee of connecting collectively collectively with your fellow treasure hunters. After all, every a success hunt is a story to be counseled and shared.

Chapter 7: Children And Young Adults

Metal detecting is not nice for adults; it's miles a interest that may be loved through people of every age. For kids and young adults, steel detecting can offer a totally specific aggregate of getting to know, exploration, and adventure. It can flip a clean walk into a treasure hunt, a history lesson proper into a fingers-on exploration, and a quiet afternoon into an thrilling discovery.

Benefits of Metal Detecting for Young Detectives:

Engaging in metal detecting can be quite beneficial for extra younger fans. It can foster a love of nature and the outside, offer a a laugh way to examine statistics and nurture trouble-fixing and studies talents. It's moreover a physical activity that can be greater interesting and engaging than conventional exercise. Most importantly, it encourages interest and a experience of wonder, turning every time out into an journey.

Choosing a Metal Detector:

When selecting a metallic detector for a child or more younger person, there are some key problems. First, the device want to be slight-weight and adjustable to deal with smaller arms and shorter palms. Second, it have to be character-high-quality, with clean settings and an clean-to-examine display. Some corporations offer metallic detectors specially designed for children, with abilties which incorporates image intention ID, easy-to-use controls, and adjustable length.

Safety and Ethics:

Before younger detectorists start their treasure-searching adventure, it is vital they have a look at the basics of protection and ethics. This includes understanding in which it's miles prison and stable to discover, commonly searching for permission while wanted, respecting nature, and schooling the 'no hint' principle—leaving the location as it grow to be placed.

Making It Educational:

Metal detecting offers extraordinary mastering opportunities. Each find can spark a research undertaking into the item's information and use. Detectorists can find out approximately precise types of metals, how a metal detector works, and the manner to apply maps and distinct belongings to become aware of promising places. Incorporating those academic aspects may want to make the interest even extra worthwhile.

Joining a Club:

Many metal-detecting clubs welcome younger participants, and a few even have unique packages for youngsters and teenagers. Joining a club can provide opportunities for mentorship, prepared hunts, and a community of pals who percentage the same hobby. It's a brilliant way to make new friends, test from professional detectorists, and participate in exciting sports activities.

Family Adventures:

Metal detecting may be a excellent family interest. It's an possibility for shared discovery and satisfaction, and it could reason interesting locations and create lasting reminiscences. Whether it is a journey to the seashore, a stroll in the woods, or a go to to an vintage domestic, each day journey may be an journey.

Encouraging kids and young adults to take in metal detecting can be a present that lasts a life-time. It's not just about the a laugh of unearthing treasures—it's miles about nurturing interest, fostering a love of information and the out of doors, and developing shared testimonies. Who knows, the subsequent swing of their metallic detector should discover a chunk of facts, a shiny trinket, or a modern day passion that lasts an entire life.

Metal Detecting Around the World: International Perspectives

A Global Pastime:

Metal detecting is a hobby that reaches past borders, taking on particular flavors in precise additives of the globe. Every u . S . Has its very private rich tapestry of history prepared to be decided below its soil, from the ancient coins of Europe to the gold nuggets of Australia, from Civil War relics in America to wartime remnants in Eastern Asia.

European Detecting:

Europe is a treasure trove for detectorists, with a big and sundry records that spans millennia. England is specially noteworthy due to its lengthy statistics of profession from the Romans through the Middle Ages to trendy times. Finds can range from Roman cash to medieval jewelry. Many European international places have legal guidelines and hints approximately metallic detecting, a few stricter than others. Always be aware of the neighborhood felony tips earlier than detecting.

Detecting inside the United States:

The United States offers a rich tapestry of historic activities pondered within the floor. From the Civil War artifacts inside the South to colonial relics in the Northeast, and the Gold Rush remnants within the West, possibilities for exciting exhibits abound. Laws on steel detecting variety from united states to country, so information neighborhood suggestions is important.

Australia and New Zealand:

Australia is famend for its gold nugget detecting, in particular in areas which consist of Victoria and Western Australia. The thrill of unearthing a gold nugget is remarkable. New Zealand, with its combination of Maori and European information, moreover offers captivating possibilities for detectorists.

Asia and Africa:

Asia and Africa is probably much less conventional places for metal detecting, but they're no lots much less rich in data. From historical civilizations to World War II relics, those continents provide a exceptional lens through which to view and understand the past.

Laws and Regulations:

It's essential to apprehend that metallic detecting felony tips vary notably worldwide. In a few worldwide places, like France and Greece, steel detecting is closely regulated or maybe banned, at the same time as in others, it is largely unrestricted. Always studies and apprehend the neighborhood criminal suggestions and pointers in advance than you start detecting overseas.

International Metal Detecting Events:

There are severa international metallic-detecting sports that convey fanatics together from all around the globe. These activities can include competitions, social gatherings, and agency hunts, providing an notable opportunity to percentage reviews, look at from others, and experience the worldwide metallic-detecting network.

Metal detecting spherical the sector can offer a wealth of possibilities for discovery. Each u.S.A. Offers its private precise historic narrative, recommended thru the relics and coins hidden below its surface. By records community guidelines and respecting cultural records, detectorists can make contributions to a broader appreciation and safety of our shared worldwide information.

Chapter 8: Turning A Hobby Right Proper Into A Business

Transforming Passion into Profit:

With endurance and backbone, steel detecting can evolve from a informal interest proper into a worthwhile task. For many, the fun of discovery is praise enough. Yet, the risk to earn a return on the time, attempt, and investment located into detecting need to make the interest even greater profitable.

Understanding the Value of Your Finds:

Before promoting, you want to recognize the charge of your discoveries. This includes proper identification, research, and, if essential, searching for professional opinion. As mentioned in preceding chapters, the valuation technique may additionally contain numismatists, archaeologists, or expert appraisers. Remember, historic and cultural fee can regularly outweigh cloth worth.

Selling Channels:

There are numerous avenues to promote your famous:

Online Marketplaces: Websites like eBay, Craigslist, or maybe Facebook Marketplace can be suitable options for promoting cash, relics, and special small well-knownshows. Ensure to offer great descriptions and smooth images for capability customers.

Antique Shops & Auction Houses: If you have got specifically treasured or uncommon gadgets, you would possibly bear in mind promoting them thru an vintage preserve or auction residence. They can frequently entice creditors inclined to pay a pinnacle beauty for unique gadgets.

Metal Detecting Shows and Fairs: These sports activities can provide exquisite possibilities to promote your reveals, network with one among a type detectorists, and meet functionality consumers.

Direct to Collectors: Building relationships with lenders can cause direct profits. This method can every so often offer the great returns, as creditors are often willing to pay extra for items they're obsessed on.

Business Considerations:

If your selling becomes everyday, it is essential to recollect the prison and financial elements of taking walks a business organization. This consists of preserving earnings, paying relevant taxes, and knowledge the legalities of promoting artifacts. Consulting with a enterprise representative or tax professional is particularly recommended.

Ethics and Responsibility:

As you embark on turning your hobby proper right into a industrial company, it's far important to uphold the ethics of metallic detecting. Never sell devices positioned on public lands or different areas in which the removal of artifacts is unlawful. Remember, our collective historical past is precious and ought to be respectable.

Balancing Business with Pleasure:

While promoting your reveals can offer economic rewards, it is critical now not to allow corporation overtake the pleasure of the quest. Continue to have a laugh with the fun of discovery, the outside journey, and the historic mastering that makes steel detecting this type of profitable hobby.

Transforming your steel-detecting hobby proper proper right into a enterprise corporation isn't for all of us. But if the idea of earning out of your discoveries appeals to you, it may add any other dimension to your detecting experience. Approach it with a balance of enthusiasm, obligation, and enterprise savvy, and you could find your self on the course to a worthwhile journey.

The Future of Metal Detecting: Trends and Innovations

Technological Advancements:

Metal-detecting era has made notable strides while you recall that its inception. Modern detectors are extra sensitive, discriminating, and person-exceptional than their predecessors. In the future, anticipate to look detectors with superior skills like 3-D imaging, advanced depth penetration, improved discrimination, and device mastering competencies to higher end up aware of buried devices.

Smartphone Integration:

With the proliferation of smartphones, metal detecting is set to become even more on hand. Apps are already to be had that flip your phone right right into a fundamental metallic detector, no matter the truth that their capabilities stay limited as compared to dedicated detectors. In the future, we may moreover see greater cutting-edge integration between detectors and smartphones, presenting abilities which incorporates GPS mapping, statistics logging, and probably even real-time identity of exhibits.

Underwater Exploration:

Underwater metallic detecting has lengthy been a specialised place of interest inside the hobby. Advances in waterproof technology and underwater imaging need to open up this thrilling vicinity to more detectorists. Imagine the thrill of diving in crystal clean waters, your detector essential you to a misplaced shipwreck weighted down with treasure.

Environmentally Friendly Detecting:

As our society becomes extra aware of environmental problems, so too does the metal-detecting community. Future traits also can embody more sustainable practices, together with better disposal of detected trash, selling 'go away no hint' ideas, and partnering with environmental organizations to easy up littered regions.

Greater Involvement in Archaeology and History:

Metal detectorists are more and more identified as valuable individuals to archaeological studies. Future collaborations amongst detectorists, archaeologists, and historians may moreover turn out to be greater commonplace, predominant to a extra understanding of our past.

Growth of the Metal Detecting Community:

The metal-detecting network is ready to make bigger within the future. This will propose huge golf equipment and occasions, a more on-line presence, and in addition opportunities for networking and collaboration. As the network grows, so will efforts to promote accountable detecting and safety of archaeological internet sites.

Educational Opportunities:

With developing hobby inside the hobby, we are capable of count on an boom in instructional property for steel detecting. From arms-on publications and workshops to on line tutorials and dedicated training packages, analyzing about metallic detecting becomes simpler and further reachable.

The destiny of steel detecting is full of thrilling possibilities. As technological innovation meets our enduring fascination with the beyond, the interest is poised to draw a new technology of treasure hunters. Whether it's miles the a laugh of discovery, the satisfaction of being out of doors, the experience of network, or the attraction of unearthing a piece of facts, the magnetic appeal of metallic detecting is about to preserve for decades to head again.

Getting Started: Your First Metal-Detecting Expedition

Planning Your Trip:

Your first excursion starts offevolved lengthy before you start swinging your detector. Do your studies, pick an area it's miles probably to yield interesting well-knownshows, and make certain you have got have been given permission to come upon there. Studying vintage maps and historical facts will allow you to pinpoint promising locations. Be privy to close by legal guidelines and recommendations, similarly to the ethics of metal detecting.

Pack the Essentials:

Your metallic detector is sincerely the begin. You'll moreover need a pinpointer for locating well-known, a digging tool for unearthing them, and a unearths pouch for safekeeping. Don't forget about spare batteries, a number one aid package, and climate-suitable apparel. If you are venturing a long way, bear in mind taking food and water, a compass or GPS, and possibly even a survival kit.

Master Your Machine:

Ensure that you're familiar with the operation of your steel detector in advance than commencing. Practice using special settings and learn how to interpret the alerts it offers you. Remember that each tool is one-of-a-kind, and what works splendid for one detectorist won't work for another.

Set Realistic Expectations:

It's crucial initially realistic expectancies. Not each experience will result in a hoard of Roman cash or a Viking treasure. In reality, most of your reveals will likely be an entire lot much less than excellent. But do not get disheartened. The pride of metal detecting comes as an entire lot from the quest and the outdoor experience as it does from the unearths.

Digging and Filling Holes:

Digging responsibly is essential. Always top off any holes you dig and attempt to leave the land as you locate it. This now not only preserves the surroundings however furthermore permits preserve the extraordinary popularity of detectorists and guarantees continued access to detecting web sites.

Recording and Reporting Finds:

Properly document the area and information of your finds. This allow you to find out styles and hotspots for future expeditions.

Depending on your u . S .'s laws, huge famous can also need to be said to applicable government.

Review and Learn:

After every revel in, make an effort to check your unearths, check your strategies, and bear in mind what you can do in any other case next time. This ongoing studying approach is what permits you to develop as a detectorist.

Your first metallic-detecting day experience is an interesting step into a international packed with data, journey, and the hazard to discover hidden treasures. Equipped together together with your detector, your statistics, and your enthusiasm, you are prepared to prompt on a adventure of discovery that could lead everywhere. Remember, it's not quite an lousy lot the treasures you unearth but moreover the studies you benefit, the locations you word, and the people you meet alongside the way. Happy searching!

Chapter 9: Understanding The Art Of Exploration

Exploration for precious metals is a complicated and difficult technique that requires specialized capabilities and knowledge. For their scarcity, splendor, and varied business makes use of, precious metals, together with gold, silver, platinum, palladium, iridium, ruthenium, and rhodium, are as a substitute prized. The exploration method includes finding mineral deposits and assessing their functionality for business viability because of the reality the ones metals may be located in rocks and minerals anywhere within the world. The numerous techniques completed for precious steel exploration, the problems encountered, and the possibilities for mining groups and investors will all be included hereunder on this section.

1.1 Techniques for Exploration:

Geology, geochemistry, and geophysics are used inside the look for valuable metals.

These techniques are hired to determine possibly places for mineral deposits in addition to to evaluate the quantity and extraordinary of these assets. The techniques used to discover precious metals encompass those indexed below:

1. Geology:

Geology is the observe of the composition, rocks, and minerals of the Earth. Geologists utilize their knowledge to discover feasible locations for mineral deposits. They search for geological inclinations, together with folds, faults, and intrusions, that would imply the life of precious metals. Satellite imagery, aerial pictures, and ground-based surveys are only some of the techniques that geologists use to find out regions which have numerous promise for mineral belongings. Geologists make use of drilling and sampling techniques to accumulate rock and soil samples for take a look at after figuring out a possible region.

2. Geochemistry:

The observe of the chemical make-up of minerals, rocks, and soils is known as geochemistry. To confirm the eye of treasured metals and exceptional minerals present inside the area, geochemists have a study samples. They have a study samples the usage of pretty some techniques, collectively with X-ray fluorescence, atomic absorption spectrometry, and inductively coupled plasma-mass spectrometry. Geologists can emerge as aware of the nearby mineral composition and take a look at the opportunity of economic viability the usage of the findings of these examinations.

three. Geophysics:

The take a look at of the bodily trends of the Earth, along with magnetism, gravity, and electric powered conductivity, is called geophysics. These characteristics are used by geophysicists to produce maps of the underground that may be used to find out viable mineral deposits. To map the subsurface, they lease a number of strategies, which incorporates gravity surveys, magnetic surveys, and electromagnetic surveys. Geologists can find out locations with a excessive capability for mineral belongings via way of way of the usage of these maps, which provide them vital records approximately the geological structure of the vicinity.

1.2 Challenges in Exploration

The following are a few problems mining groups come across at some level within the prospecting method:

1. Access to Land:

Exploration of precious metals frequently necessitates get proper of get right of entry to to to to land, which may be owned by way of manner of the usage of public or personal institutions. Accessing assets may be a hard and drawn-out device, that would slow down exploration efforts and lift expenses. Governments frequently vicinity excessive guidelines on mining operations, that would enhance the price of exploration.

2. Technical problems:

The exploration approach calls for elegant techniques and information, which might not be to be had everywhere. For the cause of creating new exploration strategies and improving individuals who already exist, mining groups want to have interaction in research and improvement. The rate of exploration may fit up if there can be a dearth of technological understanding and system.

three. Environmental Issues:

Mining groups are obligated to abide with the useful resource of stringent environmental requirements, and exploration efforts should have a huge have an effect on at the environment. The exploration device may be no longer on time and more expensive due to those guidelines. To reduce the environmental effect of exploration sports, mining organizations ought to invest in environmental manipulate structures and technology.

1.Three Opportunities for Mining Companies and Investors

Exploration for precious metals offers both buyers and mining groups a number of useful possibilities. Some of the possibilities embody the subsequent:

1. Discovery of New Deposits:

Exploration can result in the locating of new mineral property, that can provide mining corporations with a huge deliver of profits. New buyers can also moreover emerge as interested by mining companies' stock due to the invention of a brand new deposit.

2. Expansion of Current Deposits:

Exploration sports have the capacity to bring about the boom of modern-day mineral deposits. Mining companies can decorate the amount and price of current deposits, which could result in better output and profits, with the aid of locating new areas of mineralization.

three. Mineral Portfolio Diversification:

Exploration sports sports can also bring about the diversification of the mineral portfolios of mining organizations. Mining businesses can lower their reliance on a specific commodity and their hazard exposure via searching out severa minerals.

4. Investment Possibilities:

Both man or woman and institutional consumers have plenty to revel in the exploration of treasured metals. Investors can enjoy the viable increase within the price in their shares through the usage of making investments in mining groups which might be actively engaged in exploration.

In all, treasured metallic prospecting is a tough and complicated approach that calls for specialized skills and statistics. One of the techniques carried out inside the exploration way is geology. Other techniques embody geochemistry and geophysics. The exploration way is likewise difficult because of problems with land get proper of access to, technical troubles, and environmental issues. For mining agencies and traders, however, the exploration of treasured metals gives top notch possibilities, together with the finding of latest deposits, the increase of gift assets, the diversification of mineral portfolios, and funding possibilities.

The exploration approach can be crucial in gratifying the call for for treasured metals as it keeps to rise. Mining agencies is probably higher positioned to take advantage of the possibilities given via way of the exploration of precious metals if they put money into exploration sports activities and create new techniques. The feasible increase in the price of their belongings might be notable for buyers who're privy to the dangers and possibilities related to the exploration method.

Chapter 10: Origin And Ancient Developments

2.1 Early Development

By the surrender of the 19th century, as expertise and theories approximately power, magnetism, and the manner the 2 is probably blended grew, many engineers and scientists tried to create a device that might come across and pinpoint metal via the usage of this knowledge. This resulted inside the development of the metallic detector. It can be quite useful for whoever used the system within the mining organization to find out ore and rocks that contained metallic, and this is exactly what some miners did. Keep in mind that this became the nineteenth century while mining have grow to be very famous. The problem modified into that previous machines should most effective function to a restricted quantity due to the truth they ate up some of strength and were extraordinarily primitive.

Gustave Trouvé, a Parisian inventor, made facts in 1874 while he superior a handheld system that might locate and cast off metallic such things as bullets from inner human our our bodies, a feat that had in no way been finished earlier than. When a bullet became shot into the chest of American President James Garfield in 1881, Alexander Graham Bell changed into inspired to create a comparable device to do away with it. However, no matter the metal detector functioning well, the elimination of the bullet were unsuccessful because of the fact the president had been lying down on a spring coil bed, which careworn the metal detector.

2.2 Modern dispositions

The gift upgrades and changes of the metallic detector did now not begin until the 1920s.

Gerhard Fischer had created a radio path-finding device that allowed for extraordinarily particular navigation—type of on par with that of the Nineteen Twenties. Now the device operated perfectly, however the writer commenced to find out a few kinks and abnormalities in some areas in which the land had ore-bearing rocks. In accordance with Fisher's idea, it emerge as functionality to gather a tool that would come across metal the usage of a resonant radio frequency given off by way of the use of a search coil if a radio beam can be altered or bent thru metal.

He asked and bought the primary steel detector patent in 1952. The metal detector emerge as updated to a greater present day-day version inside the Twenties. A radio beam created via Gerhard Fischer became claimed to offer correct instructions. The system worked well, but there was one disadvantage. Gerhard located that the places of ore-bearing rocks diverse. He taken into consideration the state of affairs in the the front of him and concluded that considering that metals gave the radio beam an misguided analyzing, possibly one also can create a tool utilizing a are seeking for coil that might reecho at a radio frequency. He acquired permission for a metal detector patent five years later. Despite now not being the primary to use, Gerhard changed into the primary to get hold of a patent.

A businessman from Crawfordsville, Indiana, Shirl Herr, who to begin with requested a patent, changed into denied. In February 1924, he submitted an utility for a hand held hid metallic detector; it grow to be criminal in July 1928. Shirl helped Italian dictator Benito Mussolini retrieve the remnants of Emperor Caligula's galleys, which have been decided in August 1929 at the bottom of Lake Nemi in Italy.

For his second Antarctic tour in 1933, Admiral Richard Byrd used Herr's invention to find out subjects that were out of place via manner of earlier explorers. It modified into handiest effective all the way right down to an 8-foot intensity. One of the Polish officers connected to a unit in St Andrews, Fife, and Scotland all through early World War II delicate the format to a sensible Polish mine detector. Lieutenant Józef Stanislaw Kosacki changed into his name. For 50 years, no person knew that Kosacki had evolved the primary usable metallic detector because of the fact the appearance and improvement of the gadget had been carried out as part of a military studies assignment for the duration of the Second World War.

2.Three Beat Frequency Induction

These new tool creators added present day mind to the marketplace. Electronics with the aid of White In the Fifties, Oregon began after growing a gadget referred to as the Ore master Geiger Counter. Another well-known pioneering detector technician changed into Charles Garett. He invented the beat frequency oscillator (BFO) tool.

In the Nineteen Fifties and 1960s, transistor era allowed for the development of complicated circuitry in metal detectors, which allowed for the cut price in size and weight at the equal time as although walking on small battery packs. Companies sprouted up during Britain and the us due to the upward thrust in call for for gadgets. As we're properly conscious while a conductor moves in near proximity to a magnet, an electric powered modern-day-day is produced. For the Beat Frequency Induction to paintings, the detector coil want to moreover be transferring, however exceptional if the coronary heart beat is an electric powered powered EMF rather than a magnetic EMF.

2.Four Refinements

The new variations are absolutely excessive-tech and encompass covered circuits, permitting the patron to specify limits collectively with discrimination, threshold amount, sensitivity, song pace, and notch filters and reserve them only for destiny usage. The devices are deeper-in search of, lighter, use masses much less battery energy, and differentiate better than they did around ten years inside the past. Modern detectors encompass cutting-edge, substantially used wi-fi era for the headphones in addition to to hyperlink to Bluetooth and Wi-Fi networks. To preserve song of places which have been searched or the locations of products recovered, a few people rent a GPS detector. To improve the effect, others can link to clever apps.

2.Five Discriminators

The development of an adjustable induction system modified into the first-class improvement in steel detector generation. It had coils that an electromagnet used to reveal. One of the coils talents as an RF transmitter, at the same time as the opposite as a receiver. These functions can from time to time be switched to a few or one hundred KHz. Due to the eddy currents within the steel, a sign is seen at the same time as it is in its assignment. Because all metals have a superb section reaction at the same time as exposed to converting contemporary, longer current (low frequency) penetrates the soil more deeply and allows detectors to distinguish among metals and non-metals. Silver and copper have excessive conductivity, in assessment to shorter currents (higher frequency), which penetrate the soil lots less and feature gold and iron as their conductivity goals. Longer currents are tactfully diffused on the interface of soil mineralization.

Metal detectors can be created through deciding on out or separating devices if you want to select out preferred gadgets at the same time as ignoring undesired ones. Even with the distinguishers, it have come to be although hard to keep away from undesired items due to the fact a few have common reactions (consisting of gold and tinfoil), usually within the form of an alloy. Therefore, improperly changing out metals extended the possibility of lacking a massive find out. The distinguishers furthermore have the disadvantage of creating the gadgets' sensitivity less sensitive.

2.6 New coil designs

Modern techniques had been tried thru the coil's designers. Two equal coils were stacked, one on pinnacle of the opposite within the real induction coil. With the help of two D-original coils prepared back to over again from a circle, Range Electronics created a completely unique method. The Nineteen Seventies observed the majority of use for this gadget. The targeted and D kind (additionally called large scan in next years) has a large following. The improvement of steel detectors, which is probably capable of mitigate the results of ground mineralization, was each different significant improvement. Even if it wasn't terrific, this provided extra depth. Compared to the preceding ones, it truly works better at low frequencies; human beings with frequencies between 3 and 20 kHz produced the notable consequences.

In the Seventies, switches have been protected with detectors, forcing clients to choose each the differentiate mode or the non-distinguish opportunity. Later, modifications were made to permit for the digital switching of every modes. The movement detector, which continuously examined and stabilized the facts mineralization, can also moreover ultimately be derived from the invention of the induction stability detector.

2.7 Pulse triggering

The pulse induction approach, that is every different sort of metallic detecting, have come to be being explored by using manner of creators. The pulse inductor (pi) essentially affected the soil with a quite robust transitory modern from a are searching for for coil, in region of the beat frequency oscillator or the induction balancing machines, which each used a comparable abnormal current at low frequency. Since there has been no metallic gift, the floor maintained an same frequency till it was decreased to zero volts, at which issue it could be properly limited. However, if there has been metallic gift while the device fired, a tiny swirl contemporary might be injected into it, amplifying the immediately of recorded present day decline. These minute modifications in time have been made feasible via manner of advances in electronics, which furthermore made it possible to degree gadgets efficaciously and come across the presence of metals at a fantastic distance.

It modified into now possible to discover rings and other rings beneath the carefully mineralized black sand manner to the maximum critical gain of these new gadgets, which modified into that they were by manner of and large water-evidence in opposition to the capabilities of mineralization. After incorporating laptop manage, a sophisticated induction sensor is contained in a virtual sign processor. One of the advantages of the "PI detector" is its potential to "punch thru" dense mineral fields; sometimes, the dense mineral content material fabric material makes the PI detector perform greater effectively. A "PI" detector isn't always negatively affected, whilst a "VLF" detector is.

Chapter 11: Guide To Metal Detectors

3.1 What are metal Detectors?

A metal detector is a device used to discover metallic, or the occurrence or life of metal nearby, as its name indicates. They are mainly useful for finding metals that can be hid interior gadgets or in probable buried gadgets like cash, treasure, and so on. Metal detectors are typically hand-held devices with a sensor probe this is swept over the ground or other places in which the operator feels there can be metals present. The tone of the headphones adjustments because of the reality the sensor enters the steel vicinity, or in a few fashions, the indicator needle moves.

The way the tool operates is that it step by step raises the tone, which extra actually shows distance. As the metallic detector techniques the goal metal, the needle rises. The "stroll-through" steel detector is each different sort of steel detector you is probably acquainted with. It is desk bound and used in protection screening and searching at key access elements like airports, jails, and courthouses to locate and come across weapons which have been hid on a person's body without absolutely subjecting them to a whole body attempting to find.

There are diverse styles of steel detectors, along with business, pulse induction, beat-frequency oscillation, and rather low-frequency metal detectors, which we can have a observe as we skip alongside on this ebook.

The only form of metal detector uses an oscillator to create an exchange modern-day-day that flows via a coil, growing an alternating magnetic challenge. Eddy currents will unavoidably form or be generated (inductive sensor) in any steel this is nearby the coil and can behavior strength. Consequently, its magnetic vicinity is produced. The resulting exchange round and within the magnetic difficulty want to finally be detected and quantified, assuming every other coil is then used to degree this magnetic area (that is, a magnetometer). Although it looks like plenty, we can be looking at this later in the ebook.

When they had been first created in 1960, business enterprise detectors have been broadly talking hired in the mining of minerals and unique commercial packages. Land mines, weaponry like knives and guns, geophysical mining, archaeology, and treasure searching can all be completed the usage of this era. Occasionally, detectors are used to find remote places matters in meals similarly to within the advent area to discover pipes and wires hidden in walls and floors similarly to metallic strengthening bars in concrete. All metal detectors artwork in exceptional strategies; this is how the ones which might be similar paintings.

The transmitter coil, which is tied throughout the round head at the bottom of the cope with, is a cord coil utilized in metallic detectors. When an electric powered powered current circulates the coil, a magnetic field is created round it. The atoms in metals are impacted by means of manner of manner of the transferring magnetic vicinity as you bypass the steel detector all round a metal object. This modifications how electrons, which is probably tiny particles that revolve round atoms, skip. It usually tells us to ensure there may be an electric powered powered current taking walks there as well if there can be a transferring magnetic place inside the metallic. Consequently, the detector reasons an electrical contemporary-day to flow thru the steel. However, we also research some different fascinating truth from this: while energy passes over a small amount of metallic, a bit magnetic is created. The magnetic location generated via manner of the metallic detector creates a 2nd magnetic place throughout the metallic. Only the second one magnetic difficulty surrounding the metallic is picked up with the resource of the detector. The steel detectors have a circuit that includes a loudspeaker and every other coil of cord in its head, known as the receiving coil. The magnetic situation created through using the metal fragments passing through the coil at the same time as you go with the flow the detector over a metal is seen. A generator operates through passing metallic through a magnetic vicinity, which reasons metallic to conduct power. When the steel detector starts offevolved offevolved offevolved to beep or begin to thing, you have got located some issue. Additionally, the magnetic discipline across the metallic detector receives more potent as you bypass it toward the metal's supply. The reception coil receives this after it office work inside the transmitter coil. The tone receives louder the closer you get to the metal.

3.2 Uses of steel detectors

1. Archaeology metal scanners.

The very first recorded use of this type of in archaeology dates once more to 1958, at the same time as at Little Big Horn navy historian Don Rickey used one to find the fireside strains. However, the archaeologists are in competition to the use of steel detectors thru "internet web site looters" or "artifact searchers" whose acts desecrate historical locations. The vital trouble with the use of steel detectors at historic net web web sites or with the aid of enthusiasts who discover artifacts beneficial for archaeology is that the context in which the object became discovered is out of place, and no thorough analysis of its surroundings is made. An beginner metal detectorist won't understand the significance of artifacts placed out of doors of well-known websites.

2. Metal detectors for commercial use.

Industrial metal detectors are utilized by the pharmaceutical, food, beverage, apparel, material, plastics, chemical materials, lumber, mining, and packaging industries. The food business organisation faces a extreme threat of food infection from steel fragments from malfunctioning industrial device. For this motive, metal detectors are frequently hired and incorporated into the producing line. In order to test for any steel contamination (needles, damaged needles, and so forth.) in clothes, it's miles presently modern-day-day workout in the clothing or material industry to apply metallic detection after the garment has been completely stitched and earlier than it is packaged.

three. Screening for protection.

In a few places, consisting of airports and accommodations, walk-through protection instances which is probably designed to expose show display for people without in reality touching them are used to find metals like firearms, knives, and one-of-a-kind contraband.

4. Hobby.

There are many exquisite forms of leisure pursuits, along with metal detectors:

Beachcombing is the act of searching through the sand for stray cash or stones. Beach looking can be as clean or complicated as the individual chooses. The majority of ardent beachgoers are already worried approximately shore erosion and tidal changes.

Hobbyists can percent facts, show their reveals, and have a look at more about the pastime through turning into a member of metallic-detecting clubs for the duration of the United States, the United Kingdom, and Canada.

Metal detectors are regularly used to look for misplaced or abandoned objects, which incorporates useful guy-made objects like money, gold, and silver, in addition to other devices. Many metal detectors also are water resistant, permitting customers to search for submerged objects in regions with shallow water.

The intention of coin capturing is money. To find out places that could produce historically vast and unusual coins, many coin shooters conduct ancient research.

Finding valuable metals like gold, silver, and copper of their natural bureaucracy, like nuggets or flakes, is the machine of prospecting.

3.Three Maintenance of metal detectors

The safety of your metallic detector desires to come first. The device is maintained robotically to preserve it in pinnacle condition. The hazard on your business corporation might be reduced if the metal detector became used for commercial enterprise features due to the fact it'd now not be broken.

Planned maintenance and preventive protection are the two precise sorts of steel detector preservation. Although they variety significantly, each percentage the equal desires.

Maintenance this is scheduled is known as planned safety. It is primarily based totally on guidelines made via the producer concerning the bogus of precise additives. This ensures the device's clean operation.

Preventive protection, in assessment to scheduled safety, includes routine inspections of the machine to make sure it is strolling effectively and isn't always being harmed thru some aspect that would lessen its overall performance. It moreover takes into attention the behaviors which might be established across the laptop similarly to the everyday exams. For instance, developing a planned try to lower the amount of moisture the device is uncovered to.

3.Four Selecting the Perfect Metal Detector: A Comprehensive Guide

1. Selecting a required detector

Unfortunately, there is not a detector that works flawlessly on every ground. Obtaining a metallic detector have to be easy in case you use it best due to this. However, an entire lot folks need a metal detector for truely one in all a kind kinds of terrain, so that you need to cautiously pick out out one that excellent meets the majority of our desires. You can see which you have many alternatives as a capacity purchaser via manner of surfing any internet site or mag. There are many possibilities to be had, so many that even an professional detector might find out it hard. In moderate of this, the query is: How are you able to type via the market's abundance of steel detectors to find the awesome one?

You have with a purpose to attain your dreams through following the ones steps:

1. Your charge variety, your willingness to spend that amount, and accessories (including headphones, trowels, and so on) need to all be taken beneath consideration.

2. The following question you want to invite is wherein you will be doing the detection. This may be the maximum difficult item for a amateur to reflect onconsideration on, but it does now not inform you what to buy. As maximum people hunt locally, the brilliant way to address this is to appearance your community areas on a map. For example, if you stay close to a beach, you may search around there, and if you live inland and have

in no way seen the ocean, you could glide looking in parks.

3. The weight of your steel detector is a in addition thing to don't forget. A metal detector that you can not carry must not be ordered.

4. Do you pick a new or used one right now? If you get one brand-new, you'll have the assure of a producer's guarantee and, if you're lucky, some complimentary accessories, but in case you purchase one secondhand, you can pick a higher emblem. After responding to those important inquiries, you must have a look at the booklets and do not forget the detectors that correspond for your findings.

Last however no longer least, talk to someone; do no longer be afraid to ask the questions that the brochures by no means appear to cope with.

Keep in thoughts that your new detector might be your associate for a totally long term, and the success of your interest is predicated upon on you beginning with the suitable tool.

2. How to select a steel detector?

The metal detector is the most important piece of system for a newbie. If you're certain that steel detecting might be your interest. Of path, you can need to get more extras like gloves and headphones. Your steel detector, even though, is typically the primary difficulty you should select. Avoid growing a snap choice because of the truth doing so might be a grave errors. Choose the steel detector you desire to use carefully earlier than making your purchase to keep away from buying one that is beside the point for you, your desires, or your surroundings.

Before searching for one, preserve the subsequent in mind:

Assuming you stay in the desolate tract, it's far solid to expect which you might not require any scuba diving gadget. Therefore, remember what objectives is probably determined in your network in advance than shopping a metal detector. Invest in a steel detector which can choose out out coins, gold, and jewelry in case you stay in a gold-producing nation.

Learn the basics of steel detecting. It can not be emphasized enough that mastering the fundamentals is crucial earlier than starting some aspect. Consider purchasing for a smartphone at the equal time as no longer having any concept of its purpose or abilities. You might be able to use it, but you may have a completely tough time being used to it and using all of its capabilities.

Review the competencies of the various metal detector kinds. The most elegant within the marketplace are commonly Very Low Frequency (VLF) gadgets. The more cutting-edge-day-day, cordless, and waterproof pulse induction structures are in reality lots greater highly-priced.

Learn the phrases applied in metallic detecting and become familiar with the capabilities and components of your steel detector. You can also additionally select out the form of metal detector you need to buy and which sort will meet your dreams and finances in case you are aware of the settings and functions.

3. How deep can a metal detector stumble upon?

For coins-sized gadgets, a terrific metal detector can discover them as deep as six to eleven inches under the surface, however numerous of things will have an effect in this. The type of soil you're the use of for detection is one in each of them.

If the conditions are right, you may use your device to drill up to 14 or 16 inches into huge gadgets.

Several elements that could have an impact on the intensity embody:

The dimensions of the factor you're.

No rely range how deep they will be, the bigger they're, the higher your machine can be in selecting them up.

Discrimination.

It would possibly in all likelihood or won't circulate further relying on the extent of discriminating you set for your device. Low discrimination makes it impossible to even pierce via the soil's trash metallic and gain the higher targets.

The composition of the goals.

In evaluation to metals with low electric conductivity, which includes chrome steel, which makes it hard to come across at decrease depths, a few metals, together with copper, brass, and silver, make extraordinary goals and may be recognized and differentiated with out problem. How close the intention is on the subject of close by metals will have an effect on how a long manner you could dig.

The sensitivity set for your machine.

Yes, this could have a full-size effect on your steel detector. Make positive your settings are correct; in a first rate setup, even smaller matters can be decided.

4. How do the Coil Size and Type Influence the Detector Depth?

Depending on the size, exceptional, and fashion of your steel detector, you could want to put in a brand new coil. Metal detectors consist of their very non-public coils, even though. The intensity your tool can reap and its sensitivity is carefully correlated with the dimensions of your coil. In big, you'll be capable of stumble upon deeper and vice versa the bigger the coil. This has the disadvantage of creating smaller goals a lot much less complex to miss. On the opportunity hand, however the fact that your intensity is probably decreased, your metal detector will become extra touchy with a smaller coil. Small coils also are quite light in weight. As a end result, they may be less complicated to govern, this is why people generally use them while navigating difficult terrains or occurring an underground hunt. As a quit result, we endorse selecting a bigger coil in case your desires are not small. However, you ought to certainly acquire a smaller one if you are aiming for little goals like tiny gold pockets or silver. However, ensure to ask someone who is acquainted with the neighborhood soil for a coil length advice that actions a balance among intensity and sensitivity.

five. How Mineralized Soil affects the Metal Detector Depth?

Even if you realise what you're doing, mineralized soil may also nonetheless offer some worrying situations for your metal detector. As we stated within the preceding financial disaster, the quantity of metal and mineral particles within the soil generates soil mineralization, which, in turn, consequences in a magnetic have an effect on which can have an impact on your device. We do now not need to find out soil, can we? This shape of soil produces a magnetic impact that your metallic detector can select up on.

When there may be a high-quality concentration of minerals within the soil, you can experience troubles with depth, mainly in case your metallic detector isn't built to face as an awful lot as such occasions, so that it will make everything tough for you. Imagine if the place you are exploring has a immoderate iron content material. This might be very tricky and may restriction the depth and kind of the steel detector.

6. How Wet Ground impacts Detector Depth?

It is not possible to expect what sort of dust you may encounter at the same time as searching. However, the majority of human beings pick out to find wet ground. This is because it makes metals extra conductible. However, it is difficult to look beyond because of the muddy terrain. Although locating cash that have been buried turns into pretty clean whilst the earth is constantly wet, this is top notch executed after rain. Make sure your metallic detector is water-evidence in case you plan to move steel detecting inside the rain; we do not want your expensive piece of device to short-circuit. However, numerous metal detectors are without a doubt water-proof. Selecting the appropriate detector for the moist ground is essential.

7. How does the steel detector's Frequency effect its intensity?

Frequency is the form of digital waves that a device emits; within the case of steel detectors, the big style of waves which are sent to the ground enables metal detection. Longer wavelength, or low frequency, machines can acquire deeper and hit upon at decrease depths. This is because of the greater penetrating pressure of longer wavelengths. The disadvantage is that while they're remarkable at figuring out pretty conductive metals like a sliver, they may be now not in particular proper at locating little items. In assessment to their low-frequency contrary numbers, excessive-frequency detectors have quick wavelengths and relatively low depth. The benefit of this is that they may be powerful in locating gadgets which might be near the ground. Choose a multi-frequency metallic detector as the answer. Please be conscious that this shows you will be capable of use those frequencies simultaneously, no longer excellent that you can have specific frequencies on one tool.

Chapter 12: The Fundamentals Of Metal Detecting

four.1 What is metallic detecting?

Finding metal devices the use of electric powered device is referred to as metallic detecting.

There are some of advantages to metal detecting, in conjunction with the opportunity to socialise with different aficionados and the opportunity for outdoor exercise, however for all of the childlike grownups handy, finding treasure is absolutely the most appealing.

Now, what is this treasure? Well, it is some element you need it to be, something this is valuable to you. We have all heard the proverb, "One man's trash is every different guy's treasure." Go in advance and hunt for it in case you want it, and it'd supply you satisfaction to find out it; it is your treasure. Humans are born with a natural want for riches, which has been our preoccupation and ardour for the purpose that sunrise of time.

Man has continuously desired to non-public, treasure, defend, and private some issue from the begin of time. His excursions and in advance picks to have interaction in combat, devote homicide, or even positioned his life in risk all stem from his thirst for treasure. Now, while we recollect treasure, we generally think about pirates and their booty.

Simply due to how valuable artifacts might be and how often they may be determined the use of metallic detectors, metallic detecting brings you within the course of the reality of treasure than every other recreation. However, as you can quick see, it's miles going beyond handiest the financial benefit. It is the thrill of the quest, the concept of coming across and unearthing some aspect that could have lengthy beyond overlooked through man for as a great deal as a millennium. And as you retain to build up gadgets and your series expands over the years, you experience the pride of expertise that everything is yours to sit down down and understand whenever you need, in assessment to particular individuals who visit museums to view similar objects concealed inside the back of glass. Therefore, welcome to the coolest interest, whole of intrigue and childish wonder, that leads you down vintage paths of records and time and famous a myriad of uses for metals in everyday life. Some of your discoveries could probably result in economic advantage, on the same time as others may additionally beef up archaeological information or spark the discovery of a whole life. Still, others might be a bit greater unremarkable. However, they'll all relate to a story of metals that have been abused, broken, and cherished.

4.2 How to apply a metal detector?

If you've got were given your metal detector, you are organized to begin searching out hidden riches. How do you accomplish this effectively and brief? Fortunately for you, metallic detecting isn't always hard. However, like with anything, there are proper and wrong strategies to perform this, and that is what you could take a look at underneath:

1. Understand your metal Detector and its settings

Although it appears obvious, that may be a critical attention. You want to familiarise yourself on the facet of your device's settings and functionalities. Even if they'll be similar, now not all gadgets are the same. Therefore, the an awful lot less time you want to waste and the better your probabilities of locating a few factor, the greater you apprehend your tool and become familiar with its features. The super location to start is with the education manual. Study each putting and test with it for your gadget at home until you in reality understand a way to manipulate it. Additionally, do now not experience embarrassed to apply YouTube motion snap shots and instructions like this one. Despite the truth that every metallic detector is specific, the subsequent settings are not unusual to all of them:

Detection: The majority of modern-day-day detectors typically have this option that enables you to pay attention on a high quality form of item. They often embody settings fabricated from steel, earrings, and artifacts.

Sensitivity: It is most first-rate to decrease the sensitivity if the vicinity you choice to search has many underground pipelines or building tasks. The greater sensitive you are, the greater topics you could find out.

Ground Balance: While the concentrations of metals in the soil inside the direction of a seek may be small, they are able to even though reason a steel detector to move off. Simply configure the ground stability to dismiss iron under a selected diploma in case you're in search of iron and the ground seems to be overflowing with it.

Discrimination: This is a setting that instructs your metal detector to push aside precise devices which you are not seeking out. If you are in an area with masses of garbage, you will possibly configure your device to push

121

aside the word "iron." Before going to the actual deal, it is a superb concept to test your tool.

Test every parameter, together with sensitivity, discrimination, and detection, to take a look at how the signal is impacted and the manner as it have to be it detects. In order to prevent loss, we propose connecting the jewellery's stop to the stop of a rope or floss.

2. Set up your bundle deal

Although the metallic detector is the most vital device for treasure looking and metallic detecting, you will additionally require exclusive tool. Include the subsequent items for your starter package deal:

Headphones: You do not want to purchase the ones bulky over-the-ear noise-canceling headphones. But at the equal time as you discover a sign, they're the finest choice for being attentive to a beep. Any primary pair of headphones with quantity manipulate will do.

Gloves: You'll be digging generally via sand, metallic, and dirt, so that you'll want more than one sturdy gloves. Not the excellent place to reveal your sensitive pores and skin unprotected.

Coil covers: If your machine does not have already got one, reflect onconsideration on buying one. They are lots less pricey however effective at defensive your device.

Digging system: Similar to gloves, it's far crucial to have something that protects your fingers. Besides, how else are you going to break thru the soil? You can complete the mission with a tiny hand trowel or a Lesche digger.

Be aware that we propose you pick out out a first-rate tool irrespective of what making a decision to shop for. For instance, if a cheap tool is used and is derived into touch with rock or difficult soil, it can break. Alternatively, the use of reasonably-priced gloves might also additionally lead them to rupture if they come into touch with a few factor sharp, that may defeat the characteristic of the glove and result in harm.

On the other hand, avoid spending excessive quantities of cash on vain devices at the equal time as you will be saving that money or the usage of it to buy extra wonderful crucial system. There are limitless items you should buy for steel detecting, however except they'll be without a doubt critical, we advocate maintaining with the fundamental minimal.

Things like baggage and considered certainly one of a type scoops in various styles and sizes. A "pinpointer" is one crucial exception, despite the fact that. Once you've got got decided a signal, a pinpointer is a compact, moderately priced steel detector which can assist you in "pinpointing" an object. The pinpointer gives a greater unique role as soon as you have got got positioned a target and excavated a plug, permitting you to raise it up hundreds greater fast.

But is a pinpointer surely essential?

A pin-pointer may not be a few element you in reality require at the begin, however as you enhance and emerge as more worried in the sport, it's going to probably be one of the first gadgets you purchase. When you add up all of the mins the device saves throughout a session, it'd keep hours, an awesome way to make detecting more thrilling. It is common for the device to save numerous minutes at the same time as you're digging. Because of this, your probabilities are advanced, and you may have extra alerts if you dig up a find out faster.

3. Select your searching floor

The ideal places to start steel detecting are busy considering that new gadgets are constantly being introduced. The following are some of the appropriate spots for metal detecting:

1. Churchyards

2. Public parks

3. Public Schools

4. Campsites or vehicle

5. Beaches

6. Woodland

7. Private land

eight. Land surrounding motels

It is vital that you are accepted to come across metallic on a positive internet site online. If you aren't, you chance receiving trespassing fines and being barred from unique websites, so be careful not to trespass. Even countrywide parks have guidelines, and if national park officers capture you, they may capture your machine.

You must generally technique the landowner earlier than you begin seeking to get their permission. Additionally, make certain you are privy to the proprietor's belongings strains due to the reality it's far possible to dig on someone else's assets. Make sure you have got valid authorization from the close by authorities earlier than exploring land this is held via the authorities, which includes parks and different public areas.

Just be aware which you should now not visit federal parks wherein metal detecting is essentially outlawed. On the opposite hand, maximum of the seashores are appropriate for looking. Metal detecting is truly no longer authorized on personal or federally-owned beaches, as the ones are the exquisite exceptions to this rule.

4.Three What Makes a Good Location?

The preference of a hunting area can determine whether or now not a hunt is a success or unsuccessful. It's important to do your homework at the top points of interest in advance than you tour, especially in case you want to look for historic artifacts. Observables in a searching area The fine course of movement is to choose out an area that has by no means been searched earlier than, or that is typically being restocked. You may be lucky everywhere, but it is wonderful to enhance your possibilities in some of the following capability places:

Private land: It is typically almost the best choice when it comes to steel detecting, no matter being more difficult to get permission to go searching on. In essence, it's miles because personal assets stays present day and in all likelihood hasn't been searched on formerly. A amazing spot to start seeking out uncommon devices is right here.

Beaches: As previously cited, seashores are fantastic locations to look for gadgets like the ones, which includes earrings and antiquated cash. People often misplace or go away matters lying round, and you may find a startling form of riches. Because of their location, wherein storms frequently supply new topics in, they'll be regularly refilled.

Active public locations: You can continuously find out things in parks, fairgrounds, and athletic fields. There are items like coins, rings, and special stuff.

Ancient places: It may be hard to discover artifacts; you can want to be nimble and spend greater time trying to find ancient locations. They do not want to be specifically remarkable locations with plenty of because of this, but the land want to had been placed to use a totally long time ago. A park or piece of land that has been there for a very lengthy period. An wonderful region to start is the use of antique maps from earlier years. You can are also in search of Google system. Filling your plugs implies you are respectful to the land and those before them, and anywhere you dig, make certain you fill it again, or you may now not be familiar decrease lower back at the land.

4.Four Researching better are looking for grounds

The amount of studies important to find out the satisfactory looking locations would probable make this ebook far too lengthy. But for your first few hunts, we recommend that you maintain on with the standard spots. You may use the following advice to find out the right looking spots:

1. Google - The net is a splendid location, and you might be able to find out net websites in your network that would direct you to ancient web sites that have been previously lively and are exquisite locations to transport searching.

2. Local histories are published in statistics books for unique companies. Using antique maps, you could discover places that have been historically big and lively in the past.

three. Maps - Locating vintage maps of cities, regions, or ancient countries can be pretty useful on your are looking for. Places change with time; bare land can become a residing. Maps are an remarkable area to begin your are seeking.

4. Start Hunting - You have your detector, recognise the way to apply it, have offered a simple bundle deal, and are criminal to are looking for in a particular area. It's time to move looking at remaining!

four.Five Master The Right Swing

Learning the manner to nicely "swing" your detector is the number one element to do. Here are a few hints to get you going:

1. Begin with the top slightly above the floor. It should be low sufficient to permit the sign to penetrate the soil as deeply as viable without touching the ground.

2. Position the detector inside the the the front of you, feet away.

3. As you skip ahead, slowly sweep the detector in a semicircle. The detector need to be located such that it is about a foot a ways from you on each side.

four. Be positive to head right now. This implies which you can't are searching for the identical vicinity as fast as you've got moved.

five. "Low and sluggish" is the defining word you continuously pay hobby at the identical time as discussing the swing. It will take some practice to get it successfully, but within the near future, it's going to sense normal.

four.6 Digging a Plug

Stop moving and trade from a big swing to a piece circle while your detector beeps. You can greater properly determine wherein the item is due to it. When locating the satisfactory area, you may moreover need to make the detector's sensitivity more touchy. The following step is to create a "plug," this is top-rated to developing a hole. Therefore, a appropriate plug prevents the destruction of the grassroots. Here's the way to do a one-dig:

1. Create a horseshoe-common perimeter across the target region collectively at the side of your digging tool. You are not required to lessen an entire circle. Even if the item isn't as little as the bottom, just make certain you narrow the grassroots at least 3 inches deep.

2. By the usage of the digger to reveal the plug over, use the uncut detail as a hinge. Put a tiny towel right close to the hollow.

3. If you can't see the object, use a pinpointer to peer if it's in the plug or nonetheless inside the floor.

four. If you want to eliminate soil, accomplish that and region it at the towel inside the object.

five. Check the hollow another time to test if any devices are though there.

6. Reposition the plug into the hole after pushing the soil again into it.

Step on it severa instances to make certain that it is firmly in place, after which brush the grass to cast off any mats. Even although it seems excessively drawn out and hard, with workout, it receives faster. No one wants to undergo a park that is full of holes. Therefore it is essential that you leave the belongings in precise state of affairs.

4.7 Continue a Systematic Searching

Once you've got were given located your first discover, maintain shifting in advance immediately. Take steps to the facet once you've got reached the location's give up, then turn round and head inside the contrary course. Because of the mild overlap created, you may no longer forget any valuables. Despite the reality that there is not whatever incorrect with searching randomly, this sample will growth the likelihood which you might not miss some aspect.

Chapter 13: Tips And Recommendations Of Steel Detecting

For millennia, people all across the world were enthralled by means of the perception of looking for buried treasure. Metal detecting is a first rate interest with severa blessings. Metal detectorists use modern-day system to find out historic artifacts, fusing the past and gift.

Why Is Metal Detecting Such a Hit?

It is as an opportunity now not going that you'll discover a cache of hidden treasures that allows you to rework your life. Why therefore invest time and money in a hobby that surely yields scrap steel?

The thrill of steel detecting lies within the searching for and in establishing historical connections. Everything you find out grow to be interred there by the use of a person who lived a long term—likely centuries—earlier than you. There is lots potential in exploring quite unknown places since you in no manner recognize what you'll probable discover. Once you begin gathering cash or distinctive artifacts, it feels accurate to feature to your series whenever you exit.

With metallic detecting, you could skip severa time outside venture hobbies that most probably have little to do along side your day system. It can also be a shared ardour that you have interaction in together together with your closest pals or circle of relatives individuals.

Beginners' Metal Detecting Tips

Metal detecting requires a terrific deal of perseverance and training. The fantastic manner to get higher at finding treasure is to place what you have found out from the ones who've extra enjoy to paintings. The following are a few quantities of essential advice for novices in steel detecting:

5.1 Purchase a steel detector for beginners.

Beginners regularly rush to shop for detectors with superior abilties that they might not apprehend the manner to function. Modern capabilities might be a bit hard to make use of and may in the long run fail, leaving the man or woman livid and aggravated. It is commonly encouraged that beginners appoint the access-diploma version. To keep away from "toy" models, it's miles great to choose a splendid version with honest functions that you could analyze and get used to.

5.2 Invest in the top metal detector for the process.

It is maximum well-known to start off with an all-motive detector. All-motive steel detectors are suitable for locating items like coins, earrings, and lots of others., because of the fact their frequency levels from five-eight kHz. The sort of land have to be taken below consideration similarly to the form of aim. Get a water-proof or submersible metallic detector if you assume spending pretty some time in a wet surroundings. If you're thinking about stepping into gold prospecting, you need to analyze your economic state of affairs and search for specialised metallic detectors that can carry out at high frequencies, attain wonderful depths, and supply the excellent usual overall performance on mineralized earth.

5.Three Be organized to locate valuables.

Your preliminary discovery can be fascinating, but if you do now not pursue your consequences, that pleasure may additionally wane. These subjects are beside the factor, and there is probably greater. Increase your workout and familiarise yourself with the detector's capabilities so that you can apprehend your aim at the equal time as though pursuing profitable desires like gold and iron artifacts. Do not surrender.

5.Four Use a pinpointer.

A metallic detector with an included pinpointer is probably tough for novices to use to find out an object. This will be due to the search coil, the scale, the intensity, the route inside the floor, the presence of nearby metals, and the person approach. A hand held pinpointer makes it plenty less hard to locate objects correctly, which hastens the purchase of the purpose and reduces the quantity of digging required.

five.Five Completely excavate.

Dig the whole thing up. Even with a purpose ID, it may to begin with be difficult to perceive the perfect dreams. The shovel is the best tool for making distinctions. It's possible that a few proper objectives and some unwanted matters have the identical intention ID. Learn the whole thing there is to apprehend about the reactions and sounds that the intensity and conductivity of the metal causes. It could be great so that you can dig the entirety as a beginning.

5.6 Keep on digging.

If you discover one artifact, there can be greater. When you discover some aspect treasured, silver cash—which are older than maximum cash and commonly sink to the soil's backside—may be nearby. Most often, valuables are observed collectively. Use the pinpointer to appearance to the lowest whilst persevering with to dig within the identical spot. Keep looking until you encounter comparable targets.

5.7 Increase your exercising and be affected person.

The exceptional location to practice is at the actual pitch. You have to make a few mistakes, however how else will you analyze and decorate? It is generally encouraged to have a look at the training guide! Learn the whole lot there can be to recognize approximately the detector, together with its fundamental workings, its variety, and a way to super employ its skills to enhance your are searching for. To growth your abilties, watch on line how-to movies or bear in mind becoming a member of a nearby on-line metallic-detecting membership or network. As they may be saying, exercise makes perfect, so be affected character and do it regularly.

Chapter 14: Essential Tools Of The Trade

Yes, there are add-ons for metal detecting. It is truely as a whole lot as you whether or now not you require any of them. They is probably referred to as "clean naked requirements." You have alternatives: wait to see which of them you genuinely require or purchase each unmarried one that is provided. Nothing beats being well-prepared, however there's this form of element as being over-prepared. It's actual—steel detecting can be physical disturbing. If you deliver too much weight, you could no longer be capable of hunt for extremely prolonged. Starting with only some basics and walking your way up is a extraordinary idea.

Now which you with a bit of achievement have a fundamental understanding of the manner metallic detectors art work let's have a look at some of the unique equipment available, how we use them, and why. Remember that not all of these tools are important. Numerous of these tools are in particular specialized and could not also be required for the form of looking you desire to do. Like in each craft, there may be typically the "right device" for the approach.

6.1 Lesche Tool.

The Lesche device will be the first preference due to the fact the most crucial steel-detecting device or accent.

The Rambo knife of steel-detecting device, basically. A sharp, clean blade on one factor and a monster with serrations at the opportunity.

The Lesche, on the other hand, is prepared to start excavating deeply.

They are normally to be had for approximately $40 and are a extremely good investment.

6.2 The T-Handle Shovel.

A miniature spade with a 36-inch address is mounted to the T-Handle shovel. This expert digging tool is properly-named due to the T-long-established handlebars on the forestall of the shaft.

This particular device is a super choice for eliminating plugs from manicured grass. It works much like a everyday shovel and has an extended cope with that lets in you to accomplish the majority of the challenge at the same time as nonetheless popularity. Plugs may be moved without be aware, way to the clean cuts made viable with the useful resource of the pointed spade at the end.

6.Three Headphones.

It's important to have a couple of first rate, entire-ear, cupped headphones. Some of the signals you could concentrate is probably whispers which you need to pay interest very cautiously to pay interest. Additionally, you will need headphones to drown out ambient noises together with the wind, vehicles, and waves. For maximum circumstances, the sound coming out of your laptop's outdoor speaker simply may not be loud sufficient.

6.4 Handheld Pinpointers.

Since the development of the metallic detector itself, the hand-held pinpointer has been detecting's best development. Although your detector's pinpoint function is fairly particular, it could handiest provide you with a elegant concept of where to dig. It might no longer really assist you discover the item. Once the earth has been unfolded, the handheld pinpointer, that is correctly a hand held steel detector, permits with the search for the metal targets. Additionally, they'll be implemented to discover shallow goals for retrieval. While you can count on a coin in the soil to stand out like a sore thumb, they commonly aren't that easy to discover.

6.Five Tool belt and trash sack.

The trash bag is but every exceptional need. It's the top notch aspect that definitely improves how humans understand us. As if we are growing a large contribution to the surroundings, which technically we are! Therefore, do your detail by way of attaching that rubbish bag on your Bat Belt. Ensure there are not any holes in it as nicely.

The possibility that the trash dreams are concealing an awesome purpose that you may otherwise walk right with the aid of the use of manner of is one of the precise important motives to eliminate them.

A GPS, Lesche tool, and pinpointer can all in shape on your device belt.

6.6 Yard Nap.

You have to now not use a outside serviette (nap) if you need to crumble at the the front garden. The most effective way to prevent your mountain of dust from turning into a huge mess on someone's lawn is to cowl it with a bit of material, a handkerchief, or a piece of lessen-up plastic shower curtain. It's extraordinary to take a outside nap to keep the dirt in order as you do away with dust from the hole a terrific way to pinpoint a aim.

Simply pick out out up the nap and positioned the soil decrease returned inside the hollow after you've got were given finished gathering the goal. Maintaining order and cleanliness is clearly useful. The thriller to getting permission (and maintaining those you have got were given already have been given) is to deal with the assets and its owners with the most understand. The backyard nap is in reality one extra device you can use to illustrate your capability.

6.7 Root saws and clips.

It's almost tough to dig some aspect out of the earth in a few areas of the u . S . Because of the land's intense infestation with tree roots and one-of-a-kind bothersome vegetation. Most of those places are discovered inside the Northeast's forested regions. We should strongly suggest purchasing multiple gardening shears/clippers or a folding serrated blade if you take vicinity to are residing in an area with similar conditions. Try to get as an entire lot particles out of the route as you can in advance than the usage of the clippers to remove the blockading roots. But try now not to get too nuts. We do now not need to sacrifice a tree a good way to get a coin.

Once the goal has been eliminated, attempt to repair the roots to their actual position. They would probable even knit them lower again collectively. Keep a fantastic eye on what you are lowering! An unclean, gloved finger can effects be wrong for a tree root. Being without a finger inside the middle of the woods is rarely amusing and often results in a go through-associated situation. Poison ivy, poison sumac, and poison okayare greater subjects to be careful approximately.

Chapter 15: Mastering The Techniques

It's time to pay interest on a few techniques to help you refine your competencies now which you have your detector and all of those fancy devices. This chapter modified into initially intended to reputation on finding warm websites to come across and acquiring authorization, but what use wouldn't it's to get you onto those heat spots in case you did not comprehend what to do at the identical time as you arrived? So, such topics can be blanketed in the following financial ruin.

Please check to make sure you are not wearing steel-toe boots and remove any jewelry you'll be carrying in advance than we get too engrossed in specifics. You can come to be completely harassed thru the ones.

Depending on your environment, detector, and favored goal, there are numerous specific techniques to function a detector. Here, we will undergo the way to swing that trouble effectively and familiarize you with the dos and don'ts. Here is a observe of ways to turn on and function your system. Always study the manual that comes at the side of your metal detector carefully.

Turn in your metal detector and step a long way from whatever steel. To placed it some different way, get out of your vehicle earlier than you make a decision to activate your detector.

If your device has EMI cancellation skills, pick out the least noisy walking channel with the aid of adhering to the producer's tips. Keep a solid distance from strength lines, transformers, radio towers, jumbo planes, and some element else that makes your detector sound like R2-D2 on crack in case your detector isn't always able to cancel out EMI.

Conduct a floor balancing on a segment of ground that is freed from steel items. Ensure that you adhere to the manufacturer's instructions. The pleasant problem you'll find if this venture is finished improperly is disappointment.

Adjust your volume, threshold, sensitivity, and some different settings that your producer recommends.

7.1 Establish a check lawn.

Every detectorist, whether or no longer skilled or now not, wishes a take a look at lawn. A test lawn is largely a clean place of land (free of unidentified metal gadgets) wherein you've got buried numerous desires at numerous depths to test your tool's average performance.

Performance in real-worldwide instances. It's important to physical mark the places of your goals once they've been buried to save you confusion. Recording the object's intensity and description is likewise useful for future sorting out's accuracy.

You ought to bury 1 / four at 10 inches, a nickel at 8 inches, a penny at 7 inches, and a dime at 6 inches, for instance. After putting in your check garden, workout locating the desires and test with the settings to your device. Pay particular interest to your sensitivity, threshold, floor stability, and precise variables as you figure to maximize the target reaction. Additionally, play with the charge of your swing. Take be privy to the manner your detector reacts to goals which may be on the brink of not being picked up because of the truth they may be too a protracted way a protracted manner from the coil, and preserve song of such reactions. Additionally, you can bury some coins which might be perched on their aspect and going thru up and down. Since the coin's surface region is lots much less, it's miles tougher to look the significantly narrower profile of an edged coin. To in addition understand how your system reacts to numerous rubbish, attempt burying a few iron, cash adjacent to iron, and other particular portions of trash. It's a extremely good idea to practice gadget pinpointing proper right here as well.

7.2 Swing motion.

The climate, coil size, and detector type will all have an effect for your swing, however typically speaking, you want to maintain consistent touch among your coil and the ground. You will lose intensity for every inch your coil is raised above the earth. Set the shaft on your detector so that the coil is 8 to 12 inches within the the front of your toes and degree with the floor.

Most likely, the bottom of the coil of your detector is connected to a detachable skid plate. It is there to useful aid in stopping abrasions on your coil. We might also strongly advise getting a skid plate for your coil in case you've just determined out it lacks one. You're no longer maintaining your coil in near sufficient touch with the floor in case you're no longer wearing thru at least one skid plate each 365 days. Make certain to periodically take away your skid plate to easy off any dust and dirt. False indicators and depth outcomes can be due to centered volumes of mineralized soil trapped among your coil and skid plate.

7.Three Minimizing shoulder and arm fatigue.

Use your entire body to manipulate the coil whilst gripping the device, maintaining your elbow tucked in near your hip. It will appreciably reduce arm and shoulder fatigue. If you want to swing for a long term, it is probably beneficial to workout swinging together together with your non-dominant arm.

We must propose schooling in a flat place with little to no barriers. It can be quite uncomfortable or perhaps risky at the start to discover along with your non-dominant arm. Wait until you see your self detecting left-surpassed in case you idea you have been uncommon throwing left-exceeded. Although it's now not best, it can be the distinction among a half of of-day and a full-day hunt. You may even avoid a few severe continual shoulder and arm problems in the destiny by using manner of using switching palms.

7.Four Swing Pace.

Some machines have processors which is probably quicker than others. Usually, the more pricey tool can technique more ground faster. However, if you have a large open vicinity with few to no objectives and a quicker swing velocity, you will be capable of cover greater floor trying to find warm spots (areas with quite some signs and symptoms indicating large human pastime). Obviously, whipping that thing spherical at moderate tempo may not produce exceptional effects in an iron-infested website online.

Make amazing to overlap your swings no matter how slowly or speedy you're swinging that object. The maximum coverage consistent with step is supplied through overlapping your swings, which also prevents you from by means of twist of fate strolling over gadgets and narrowly missing them.

Slow down at the equal time as you're in a warm place! Make positive to overlap the ones swings as thousands as possible. If you have got one, you can want to vicinity on your smallest DD coil. It will assist you in distinguishing some of the alerts and nicely and awful desires.

Some VLF detectors react more favorably to a faster swing. These unmarried-frequency VLF gadgets' extra swing speeds can on occasion redesign a susceptible, idling sign right right into a screamed: "Dig me!" It's critical to undergo in mind that your swing velocity have to be sluggish in places with masses of goal consciousness, however the fact that those machines may also additionally additionally respond better to a faster swing. Once you've got recognized a possible purpose, you may speedy circulate the coil over the suspected goal to be able to enhance the sign.

The sort of objectives in the ground, the quantity of obstructions at the floor, and the shape of gadget you are the use of want to all be immediately correlated at the side of your swing velocity.

7.Five Gridding.

You need to grid the place at the same time as you've got discovered a excellent region that has yielded some exceptional coins, relics, nuggets, or unique riches. There are numerous unique gridding strategies, however all of them need to, in the long run, really cover each square inch of immoderate-yielding terrain. It might be as simple as searching out your footprints with the aid of strolling backward and forward in instantly lines, or it may be as correct as making use of GPS monitoring.

7.6 Don't flow slowly.

One of my desired gridding techniques on the dirt is to in fact drag my feet, leaving a completely apparent route. However, in case you are gridding a sincerely big vicinity or (a) have terrible footwear, this is not the fine alternative. If you undertake this technique and function a huge location to grid, you will, if you do no longer already, quickly have terrible shoes.

7.7 Drag a chain.

A piece of eight to ten-foot rope can be used as an possibility to gridding in the dirt with the resource of attaching some links of robust, appropriate-tremendous chain to at the least one stop. Then strong the rope's distinct surrender spherical your waist. The rope will course you as soon as you begin to flow into. A mainly dense line is probably common in the again of you with the resource of the massive chain link on the quit of the rope. It's a terrific, fingers-free approach of gridding on an excellent price range. Naturally, this technique moreover dreams some assistance from the surroundings/terrain.

Chapter 16: Identifying, Cleaning And Promoting

eight.1 Identifying the unearths

You need to first come to be aware about your gadgets in an effort to have a look at the proper strategies for cleaning and providing them. It can be pretty easy to pick out your findings or very complicated. It all is based on what you get hold of as real with your discovery to be. Joining one of the many metal-detecting communities will offer you with a solution when you have some thing and have no earthly perception of what it can be.

1. Identifying Old Coins.

To choose out older cash, you may both use some of the metallic-detecting corporations or conduct your personal take a look at. For this kind of have a study, the Internet is a exceptional useful aid. Sometimes all it takes is a brief Google searching for to discover what some of the textual content content on a coin's face method. Pay excellent hobby to dates and mint markings especially. The best indicators for ancient coin identity are often the ones. This is a fantastic manner to recognize cash, but what occurs in case you do no longer have get right of entry to to the Internet? To assist you in figuring out your coin reveals, we advise purchasing a certainly best coin e-book.

2. Identifying Old Relics.

Relics may be pretty hard to pick out out out, specially if they may be made of iron and had been buried for a very long time. They also can truely have a faint resemblance to who they as soon as have been. There can be typically while you locate artifacts that without a doubt do not belong within the vicinity, but in case you finished a bargain of research at the area wherein you're looking, you have to have an terrific notion of what you may have observed. When this occurs, we propose you to attempting to find recommendation from the metal-detecting community. There will typically be someone who can virtually find out your artifact, no matter what you've got positioned. Additionally, you can use the Internet to apprehend some of your older artifacts, specially if they may be navy relics like Civil War bullets. There are numerous first-rate assets that describe the ones varieties of artifacts. You can discover some top notch effects through trying to find "Civil War bullets" on Google.

three. Identifying Jewelry.

You have a completely strong threat of locating a few antique jewelry most of the loot you unearth. Removing a plug and recognizing a shimmer of gold in the floor under may be exciting. When that flash of gold seems to be an antique ring studded with diamonds, rubies, emeralds, or each other valuable stone, the enjoy is probably even more interesting. But how can you decide what that ring is made from? If the ones stones are real, how are you going to inform? Unbelievably, it isn't that tough to understand antique rings. A few gear can be important for the gadget to obtain achievement. For rings identification, a remarkable jeweler's loupe is critical. They are effectively to be had on-line for only some greenbacks. A magnifying glass can also be used, but it is able to no longer be as powerful. Nearly all jewelry have markings within the band.

four. Identifying Diamonds or Other Precious Stones.

It takes ability to differentiate a real diamond from a fake. Mastering this art work can take years. You are attempting to find minute particles of carbon even as you study what you consider to be a diamond. They will display up as black specks scattered across the stone. You may additionally moreover take a look at the alleged diamond up near and personal using a effective jeweler's loupe. A diamond may be properly really worth a good deal less if it has greater carbon or black spots. A diamond may be properly well worth more if it has much less carbon. The readability of the stone is impacted by those carbon flecks. If you do no longer look at any carbon, you is probably retaining a totally pricey diamond. There are some further symptoms that your treasured stones would likely or may not be actual. Diamonds set in silver jewelry are uncommon.

There are a few super methods you can decide whether or not or now not or no longer your these days determined treasure is right. You may additionally additionally locate virtual diamond testers that may decide whether or not or now not or now not or not your diamond is actual to be had inside the market. They do now not charge lots of cash. You can also have enough cash to buy a diamond tester if you are locating an entire lot of jewelry with diamonds. Another choice is a neighborhood jeweler. Although your jeweler should now not have any problem figuring out your presently determined object of jewelry, usually exercise warning spherical jewelers.

There are many dishonest people on hand, and they'll profits significantly from your lack of expertise. The jeweler ought on the way to estimate the truly properly well worth of your new acquisition. Just keep in mind that price is usually a relative concept.

eight.2 Cleaning Your Finds

Cleaning is non-compulsory. The question is that! In unusual times, washing your treasures ought to truely lessen their nicely really worth. Cleaning your new treasure might be now not an top notch concept if there may be any uncertainty for your mind. This is mainly real for vintage cash with appealing natural patinas. Compared to in case you had wiped clean them, they're masses more precious as-is. The equal might be stated for any historical artifacts you may stumble upon. Additionally, there may be instances whilst you want to clean a product so you can sell or showcase it as it should be. Here are some of the first-rate cleansing techniques for a number of the treasured devices you'll be uncovering.

1. Clad Coins Cleaning.

You will in the long run have a big collection of clad cash. You have to even need to deposit they all at the financial institution. Because you do not have to fear about unfavorable those cash, cleansing clad is quite clean. They have no other monetary sincerely without a doubt well worth past their cutting-edge-day face fee. The first rate technique to thoroughly easy the ones sorts of cash is with a rock tumbler.

2. Cleaning Valuable Coins.

A few elements determine how a good buy older cash are properly nicely really worth. The good sized form of equal cash produced in addition to the coin's gift state of safety. You is probably doing something simply incorrect in case you manipulate to ruin the coin at the identical time as cleansing it. You need to leave coin cleaning to the experts if you have an older coin that needs to be wiped smooth. There are corporations that offer professional coin cleansing and grading services.

There are a few attempted-and-proper techniques that artwork quite properly if you want to clean an older coin that isn't always particularly precious yourself.

The approach for slowly cleaning coins.

Soak your coin in olive oil in case you don't mind prepared a bit bit (every now and then 2-3 months). Even the dirtiest coin may be wiped smooth with this technique with out obvious harm. The drawback of this method is that it could take a very prolonged duration. Stick with this technique if you have the staying power. If you are not able to wait, don't forget some of these other powerful cleansing techniques.

Options for faster coin cleansing.

On rusted-out, vintage metals, toothpaste is a miracle worker. Apply a tiny little little bit of toothpaste together together with your fingertips to the coin's floor. To assist in dislodging some of the dirt at the coin, you can also use a easy-bristled toothbrush. Continue till the coin has a pleasing look. Another smart technique makes use of a small amount of technological expertise. Use a few wet tinfoil to rub the antique coin. This cleans the coin's floor and produces a slight electrolysis effect. If you pick out out, you can even truely electrify the area. Using baking soda, vinegar, and lemon juice to clean cash has additionally been a achievement for a few people. They can all be applied on my own or as a few shape of peculiar aggregate created with the useful resource of using a mad scientist. Just be cautious whilst combining substances. You may be setting yourself up for all styles of issues.

3. Jewelry cleaning.

There are some topics you can use earlier than taking an vintage piece of jewellery to a jeweler, who can constantly make it gleam like new. The earrings seems as unique as new after being wiped smooth with toothpaste. Simply comply with toothpaste to the jewellery's floor. Polish the object in question with toothpaste and a slight fabric. Once the jewellery is nicely first-rate, rinse and repeat. With the exception of reasonably-priced jewelry, that is effective with pretty masses any sort of metal used to make earrings.

Because it's far rubbish, cheap jewelry tarnishes and turns into filthy brief. There isn't hundreds you can do to this sort of earrings to make it even remotely attractive.

four. Precious stone cleansing.

Diamonds, in particular, will be predisposed to get a bit dirty over the years. The pores and pores and pores and skin's oils and greases adhere to the floor of valuable stones and bring a number of problems that lessen the stone's attraction. The stones will become extra attractive after being wiped clean in addition to regaining their true luster. Before you begin attacking your valuable stones, there are a few matters to think about. Any damage you cause is definitely your fault. Alternatively, you would likely certainly as without issue knock the stone or stones out of the jewellery item. They ought to probably drop into the sink, causing you to lose your treasured item all over again. If you are uncertain about cleaning rings containing stones, have a pro manage it. Here is what you need to do in case you want to finish this mission at domestic.

Water this is heated is the primary problem. Warm water might not make treasured stones sparkle and glitter, but it will assist dissolve any oil or grease that could have built up at the stone's floor.

You can mix some drops of ammonia in with the water, a small amount of liquid earrings cleansing, or Ivory brand dish soap. This combination aids in cleaning off any debris that might have collected at the floor and base of your just sold valuable stones.

5. Cleaning Relics.

Relic cleansing techniques largely rely upon the cloth that the relic is built of. You could probably determine now not to smooth the artifact in a few occasions. It may have greater truly well worth in its contemporary-day situation. It's time to enter a probably unstable region for the ones of you who're decided to clean up those rusty antique relics. Priorities come first.

We'll be discussing every water and energy. These items do now not skip alongside nicely. You can also need to revel in a lifestyles-changing shock from a single mistakes. You want to hold at your non-public threat due to the fact it may also result in loss of life. Discuss electrolysis now.

Chapter 17: From Backyards To Historical Sites

There is a plethora of capability hiding spots for treasure. You can locate loads, whether or no longer you're seeking out antique artifacts, cash, jewelry, or gold nuggets. The odds of creating a giant discovery beautify in case you click on at the quality are seeking link. Locations with excessive stages of foot site visitors and historic web sites that have now not been decided through the majority of detectorists commonly generally generally tend to yield the first-rate outcomes. The most effective disadvantage is that, besides a few beaches, all the first-class locations to appearance are on non-public belongings.

Trespassing, as you well apprehend, remains trespassing, steel detector or no metal detector. If you've got any doubts about a way to proceed, be sure to take a look at the rules. It's every now and then truely worth the hassle, as getting stuck might also moreover bring about fines and likely the shortage of your metal detector. Obtaining approval from personal landowners is not continuously simple, irrespective of the fact that it's also plausible. The first step is deciding on the proper looking region, wherein you may have a higher hazard of bagging a trophy seize.

9.1 Places to look as a amateur

These are the primary locations a newbie treasure hunter must look. Always endure in mind to update your plugs and use the litter. It's vital to move away a are attempting to find net site in better form than you determined it in.

1. Your very private Backyard.

Even if you're now not specifically interested in finding treasure for your very personal out of doors, it's miles a notable spot to practice doing an prolonged, even sweep with a metal detector. You can discover how well the detector works through hiding special metals in their lawn (a check lawn). If you're fortunate, you might unearth coins, rings, or a few element truly right away. The first-rate factor is that you do now not should ask simply absolutely everyone's permission to do a little element in your very own outdoor.

In case you have not already, here's a rundown of metallic detector classes broken down through price point for scuba divers.

2. Farmland owned by means of manner of the family.

When you have got were given exhausted the opportunities for your very personal outside, the following logical step is to inquire about finding out the belongings of buddies and associate and kids. Yards are often modest, so that you may not flip up any very interesting artifacts. If you suspect your buddies is probably interested by your belief, you can ask them.

three. Sports Stadiums.

People's attention has a tendency to wander at wearing activities and specific venues, making it easy for free alternate and distinct valuables to fall to the floor. The simplest seize is that almost all of detectorists already apprehend this. As a result, you want to hurry to the stadium at the identical time as the game ends. Sports arenas are normally privately owned and so require authorization earlier than you put foot inner.

four. Close-thru Parks.

The golden rule of famous metallic-detecting spots is that extra people equals more out of place treasure. Finding cash or earrings in parks is a popular hobby. The out of place gadgets are not unusual at outdoor sports sports like picnics, video video games, or perhaps walks with the dog. After the truth, it's miles better to search for a public meeting. To effectively excursion public parks, you ought to first accumulate the correct allows. You would possibly get a few abnormal appears in case you do this.

five. Beaches.

Water, slippery solar cream, and lively beach video video games all make a contribution to making it easy to misplace valuables. Every day, humans lose earrings, coins, and special valuables on the beach. Beaches are ideal due to the fact the waves there regularly wash up antiquities. A allow is not required to visit a public seashore, but you can want to collect written authorization to go to a personal seashore. The important downside is that the seashore is a famous tour spot for steel detectorists. Going past due on busy days or after storms will growth your odds of finding a few issue of charge.

6. Churches.

While secular systems arise and bypass down, maximum church homes have stood the test of time. That's why it is one of the incredible locations to find out every present day-day rings and vintage artifacts. Searching for a cemetery with out permission is specially inappropriate, so as an alternative, attention on regions in which people are in all likelihood to be walking or sitting, along with round huge trees and pavements.

7. Battlefields.

Numerous battlefields from the American Civil War even though exist, and the identical is right of ancient and modern battlefields in Europe. These locations are wonderful alternatives if you're looking for ancient stuff. The simplest capture is that the number one battlegrounds also can have already been scanned, but you never apprehend. Find the nearby archaeology department and allow them to apprehend if you stumble over a few element of historic importance. Everyone can be relieved to pay attention about your locating.

eight. Woodland.

Wooded regions frequently offer trails for strolling puppies, walking, biking, and trekking. It's feasible that plenty of things might be out of region alongside the road; despite the fact that, if you're lucky, you'll in all likelihood moreover come upon a few historic artifacts. The tracks may additionally moreover furthermore have shifted thru time, but we can be effective that humans have normally traveled that route. Even even though forests are taken into consideration public assets, you still need to get permission to enter them.

nine. Fields.

It's not surprising that the majority of the maximum big steel-detecting unearths had been made in open areas. Agricultural terrain that has been used for farming isn't going to were very well explored. Plowed fields make excellent looking grounds. When discarded soil reaches the ground, it brings some thing modified into buried in it with it. Research is useful due to the reality that it's miles possible to find out surprises in any vicinity.

You might also moreover have town hall meetings in a nearby farmhouse or barn or perhaps an deserted area. An incredible place to start is with the library; there, you can locate vintage maps, newspapers, and splendid archival materials, and you will in all likelihood moreover communicate to lengthy-time citizens. To are in search of private property, you'll, of route, require authorization. Big troubles ought to get up if you're given the green moderate. Any vicinity in which humans are performing is a possible location for buried riches.

nine.2 Gaining Permissions: Navigating Legalities and Etiquette

There are some valid reasons why nearly all metallic detecting takes location on non-public assets.

Antiquities prison suggestions and particular authorities rules are the number one motive, which may additionally come as a surprise to you.

You understand, in case you discover some detail fantastic on public grounds, there's a strong chance you can no longer be capable of keep it and a terrific better danger you may want to pay a rate for, say, harming an endangered snail's habitat.

The public spaces which have been as soon as taken into consideration honest undertaking have been crushed to loss of life over the preceding half-century to six a long time, which brings us to the second cause why non-public belongings is favored. Those public places that used to attract human beings are truly simply barren. You have a better danger of locating keepers and keeping finds in case you're on private land.

You'll must knock on some doorways and convince a stranger to assist you to gather desires from their well-saved garden except you very very very own a big tract of land that is probably ideal for detecting.

In this section, you'll discover ways to get your hands on the goods. It's never easy. If you follow a number of these important techniques, even though, you have to listen "pass in advance" on kind of 80% of your efforts.

1. Be selective.

Improve your possibilities via exploring your options. Look at the real belongings, whether or not or now not you are discovering it on line or just driving approximately.

Take be conscious: If you're doing this on line, you have to use every satellite pix and road view.

What's the circumstance of the grounds? Is it simply me, or does the the the front backyard seem like some form of hillbilly recycling center? Do you believe you studied the weeds are too excessive to swing through? Are there symptoms everywhere that say "No Trespassing" and "Don't mind the canine, watch out for the proprietor"? The terrain need to also be taken into consideration. If you are not in pinnacle physical state of affairs and do now not revel in like mountain climbing hard terrain all day, do now not trouble. Keep it in thoughts and located it away for a wet day if it's far historical but may not appear to be the first rate vicinity to type thru it right now. In a bind, you may continuously fall again on the ones possessions.

2. Internet Research

This detail must make you uncomfortable, however pay attention me out. Find out who the prison owner of a piece of land is with the assist of websites like ReportAllUSA.Com and mobile apps like LandGlide. Do a Google are looking for with the proprietor's call and address after you've got were given identified them. Adding an deal with to your are seeking normally yields extra correct results. However, this could variety counting on how good sized the call you're the use of is. Facebook profiles, businesses, information memories, honors, economic contributions, organizations, memberships, or maybe criminal facts are all with out trouble available.

3. YouTube Approach

You want to truly upload your motion pix to a social media platform like YouTube for the YouTube approach to succeed. A large-rate variety Hollywood film isn't required. In fact, the sizeable majority of modern smartphones come prepared with video recording and uploading capabilities to YouTube.

4. Article Approach

In our enjoy, the "article technique" has been one of the simplest strategies of acquiring approval. Writing articles for treasure-hunting journals, blogs, and ancient society newsletters is a incredible way to gain get entry to, but it does require electricity of mind. Western & Eastern Treasures, American Digger, Lost Treasure, Gold Prospector, Coins, and Coin World are only a handful of the diverse well-known journals committed to the pursuits of treasure searching, prospecting, and coin amassing. The magazine racks of your network book shop are in all likelihood to house the large majority of these courses inside the "interest" section.

Metal-detecting lovers can also percentage their memories and discoveries on numerous online message forums and community blogs. A few examples encompass Treasurenet.Com, Findmall.Com, and MetalDetectingForum.Com.

After you have written your put up, you want to select wherein to post it. You can post your article for assessment to maximum treasure-looking, prospecting, and coin-gathering periodicals via following the enclosed commands and speak to information. If your article is chosen for guide, a representative from the magazine gets in contact with you to exercising the data in a short settlement. You can count on to earn a truthful wage on your efforts most of the time. Creating an account and signing up is the most effective step required to position up your writing in online forums and blogs. After signing up, you'll have unlimited get right of entry to to characteristic as many articles and images as you need. No one is paid to jot down for on-line speak companies, however some blogs make cash from advertising. Once you've got got something to reveal on your writing talents and a functionality spot for a destiny function story, you could use the object as leverage to get permission.

You can introduce yourself and offer an reason behind which you're searching into the county's records to provide a bit of writing in your mag with the resource of pronouncing some thing like, "I'm doing some studies on the county's statistics so that it will provide better statistics approximately the area's past."